Tissue Repair and Reconstruction

Series Editors

Andy H. Choi, Carlingford, NSW, Australia

Besim Ben-Nissan, Sydney, NSW, Australia

SpringerBriefs in Tissue Repair and Reconstruction provides a unique perspective and in-depth insights into the latest advances and innovations contributing to improved and better treatments for patients with damaged soft and hard tissues as a result of diseases, trauma, and implantations. The book series consists of volumes that offer biomedical researchers better insights into the advancements of biomaterials science and their translation from the laboratory to a clinical setting. Similarly, the series provides information to surgeons and medical practitioners on novel ideas in biomedical science and engineering on top of disseminating new ideas and know-hows in diagnostics and treatment options for patients from head to toe.

The series will cover a number of key topics:

Fundamental Concepts and Surface Modifications: The topic will provide detailed information on the discovery and advancements of biomaterials surface modification approaches and their use within the human body in a safe manner and without provoking any negative tissue response.

Computational Simulations and Biomechanics: Anatomically accurate computational models are being in all fields of medicine particularly in orthopedics and dentistry to reveal the biomechanical functions and behaviors of bones and joints when damaged, diseased, and in the health state. They also contribute to our understanding during the design and applications of implants and prosthetics subjected to functional loadings and movements.

Surgical Advances and Treatment Options: Discusses how surgical techniques are revolutionized by our deeper understanding into biomaterials science and tissue engineering. The section also focuses on the latest innovations and surgical advancements currently being used to treat patients with damaged tissues.

Post-Operative Treatment and Rehabilitation Engineering: Expands the independence and functionality of the patient after surgery while at the same time reducing the chance of complications such as wound infections and dislocations. Advances in technologies are creating new opportunities in how physiotherapy rehabilitations are delivered.

Andy H. Choi · Besim Ben-Nissan

Hydrogel for Biomedical Applications

3D/4D Printing, Self-Healing, Microrobots, and Nanogenerators

 Springer

Andy H. Choi
University of Technology Sydney
Sydney, NSW, Australia

Besim Ben-Nissan
University of Technology Sydney
Sydney, NSW, Australia

ISSN 2731-9180   ISSN 2731-9199 (electronic)
Tissue Repair and Reconstruction
ISBN 978-981-97-1729-3   ISBN 978-981-97-1730-9 (eBook)
https://doi.org/10.1007/978-981-97-1730-9

© The Editor(s) (if applicable) and The Author(s), under exclusive license to Springer Nature
Singapore Pte Ltd. 2024

This work is subject to copyright. All rights are solely and exclusively licensed by the Publisher, whether
the whole or part of the material is concerned, specifically the rights of translation, reprinting, reuse
of illustrations, recitation, broadcasting, reproduction on microfilms or in any other physical way, and
transmission or information storage and retrieval, electronic adaptation, computer software, or by similar
or dissimilar methodology now known or hereafter developed.
The use of general descriptive names, registered names, trademarks, service marks, etc. in this publication
does not imply, even in the absence of a specific statement, that such names are exempt from the relevant
protective laws and regulations and therefore free for general use.
The publisher, the authors, and the editors are safe to assume that the advice and information in this book
are believed to be true and accurate at the date of publication. Neither the publisher nor the authors or
the editors give a warranty, expressed or implied, with respect to the material contained herein or for any
errors or omissions that may have been made. The publisher remains neutral with regard to jurisdictional
claims in published maps and institutional affiliations.

This Springer imprint is published by the registered company Springer Nature Singapore Pte Ltd.
The registered company address is: 152 Beach Road, #21-01/04 Gateway East, Singapore 189721,
Singapore

Paper in this product is recyclable.

Preface

The utilizations of hydrogels in modern medicine are increasing especially in areas such as regenerative medicine and tissue engineering. Hydrogels have been demonstrated to be a promising scaffolding material due to their physical, mechanical, and chemical properties as well as the capacity to entrap cells within the material itself. Most importantly, the use of 3D/4D printing and bioprinting enables the fabrication of tissue scaffolds that precisely match the defective area. The printing of hydrogel-based scaffolds is appealing as they can imitate the cellular environments found in vivo where features such as mechanics and high-water content are similar to the natural extracellular matrix.

3D bioprinting offers numerous advantages such as high-resolution cell deposition and accurate control of cell distribution. Bioprinting, distinct from conventional 3D printing approaches, requires a different technical methodology that is compatible with depositing living cells. Deposition approaches such as laser-induced forward transfer and extrusion-based printing have been studied for the biofabrication of hydrogels as a carrier for cells and/or bioactive compounds, and they require very specific requirements to cater for the properties of the hydrogel-based bioinks used such as their post-curing rate and rheology to achieve reliable fabrication of 3D structures and to ensure cell viability.

Biomaterials with features that allow them to heal themselves when damaged by mechanical, thermal, or other mechanisms and restore their original sets of properties autonomously and automatically without any external intervention have been the pinnacle of regenerative medicine. Self-healing hydrogels are of particular interest, especially in tissue engineering. Furthermore, injectable hydrogels that can self-heal are gradually explored and applied as inks in 3D extrusion-based (bio)printing, as they can be fluidized temporarily under shear stress and regain their mechanical characteristics and original structure once the applied stress is released. Smart or stimuli-responsive hydrogels can also be used to create dynamic 3D-printed tissue constructs that can change its shape when exposed to an external stimulus such as temperature, pH, magnetic field, or moisture. The evolution of 3D printing technology along with our improved understanding into biological systems have created new opportunities and prompted the exploration of stimuli-responsive hydrogels in 4D printing. 4D printing can potentially benefit several areas in the medical arena such as tissue regeneration and drug delivery.

The developments of hydrogel micromotors and microrobots have shown significant potential in areas such as drug delivery. The biggest difference between micromotors and microrobots compared to conventional micro- and nanodelivery systems is their ability to self-propel. Recent advances into the research and development of micromotors and microrobots have made them promising tools in addressing several biomedical challenges due to their unique characteristics such as high towing force and cargo loading and fast motion. Presently, considerable efforts have been made towards the development of a self-powered, controllable, and on-demand drug delivery system driven by triboelectric nanogenerators. Their energy harvesting and self-powering properties make triboelectric nanogenerators a potential candidate for next-generation targeted drug delivery.

It is envisaged that this book will provide a fundamental insight into self-healing and stimulus-responsive hydrogels fabricated using 3D/4D (bio)printing for tissue repair. The book will also discuss the in vitro and in vivo responses of hydrogels crosslinked using enzymes and copper-free "click" chemistry. Furthermore, it will also examine the future of drug or therapeutic delivery using microrobotics and self-powered devices based on triboelectric nanogenerators. Lastly, I would like to acknowledge the people at Springer, especially Dr. Ramesh Nath Premnath, Bhagyalakkshme Sreenivasan, and Ramamoorthy Rajangam, for their help throughout the publication process.

Sydney, Australia   Andy H. Choi

Contents

Chapter 1
Brief Introduction and Various Crosslinking Approaches

Numerous descriptions have been conceived to describe hydrogels and the one that has been widely accepted is that it is a crosslinked, water-swollen three-dimensional polymeric network composed of synthetic or natural hydrophilic polymers synthesized using one or more monomers via simple reaction. Several important classes of hydrogels have been created based on the way they were prepared and they can be categorized as interpenetrating network (IPN) (the amalgamation of two polymers in the form of a network and at least one of which is crosslinked and/or produced near the other [1]), semi-IPN (can be formed in the absence of a crosslinking mechanism to generate a network of embedded linear polymers entrapped within the original hydrogel [2]), homopolymeric (crosslinked polymeric networks obtained from one kind of monomeric unit [3]), and copolymeric hydrogels (consisted of two different monomers and at least one monomer is hydrophilic in nature [3]).

Hydrogels have also been thought of as a polymer that demonstrates the capacity of being insoluble in water but able to swell and hold a vast quantity of water within its structure. It has been suggested that the hydrophilic functional groups attached to the polymeric backbone of hydrogels were the mechanism behind how it is able to absorb water. In addition, crosslinking among network chains provides the hydrogels resistance to dissolution [4]. Crosslinking can be carried out using several approaches including chemical crosslinkers, pH, photocrosslinking, or adjustment in temperature [5].

The ability to swell in the presence of water, and vice versa, shrink in the absence of water has been recognized as one of the most characteristic features of hydrogel [6]. Hydrophilicity or the nature of polymer chains and the crosslinking density governs the amount of swelling. According to Park and Park, the swollen network of the hydrogel will collapse during drying because of the high surface tension of water and for that reason, dried hydrogel is much smaller in size in comparison with a hydrogel that is swollen in water [6]. Most importantly, hydrogels can maintain its overall shape through the shrinkage and swelling process.

© The Author(s), under exclusive license to Springer Nature Singapore Pte Ltd. 2024
A. H. Choi and B. Ben-Nissan, *Hydrogel for Biomedical Applications*, Tissue Repair
and Reconstruction, https://doi.org/10.1007/978-981-97-1730-9_1

In general, the infusion of nutrients into the hydrogel as well as the cellular products coming out can be regulated by the character of the water [7]. Hydration will take place with the most polar and hydrophilic groups as soon as the first water molecules enter the hydrogel matrix once a dry hydrogel starts to absorb water. This is referred to as primary bound water. Secondary bound water or hydrophobically bound water is the result of the swelling of the network due to the hydration of the polar groups causing the hydrophobic groups to be exposed and interact with the water molecules. The term total bound water is frequently applied to describe the combination of both primary and secondary bound water [7]. A hydrogel is referred to as superabsorbent if the amount of water is greater than 95% of its total weight. Typically, a minimum of 20% of the total weight of the hydrogel is water [6].

An equilibrium swelling level or a maximum degree of swelling will be reached by the hydrogel [7, 8]. In general, this occurs when a balance between the cohesive forces (which resist the expansion of the hydrogel) exerted by the polymer strands within the hydrogel and the osmotic driving forces that promote the ingress of biological fluids or water into the matrix of the hydrophilic hydrogel is achieved [9]. A larger quantity of water can be absorbed by the hydrogel if it is fabricated using a polymer that exhibits a greater extent of hydrophilicity. Furthermore, a reduction in the magnitude of gel swelling can be achieved simply by increasing the quantity of crosslinking in the hydrogel [9].

A study has postulated that a hydrogel with adequate mechanical attributes and large swelling capacity can be produced through appropriate selection of the co-monomers and the intensity of the crosslinking agents [8]. This is based on the notion that a reduction in the mechanical properties is associated with large swelling. According to Hoffman, the hydrogel will begin to disintegrate and dissolve as the network swells at a rate governed by its composition only if the crosslinks or network chains are degradable [7].

Hydrogels are appealing for many biomedical applications including tissue engineering and regenerative medicine due to their unique biocompatibility as well as high-water content. Furthermore, it has also been suggested that this rise in interest was partly due to the similarities between hydrogel and extracellular matrix in terms of composition and innate structure in addition to their extensive framework for cellular proliferation and survival [10].

Once implanted, hydrogels can confront complex geometries and are ideal in applications where minimal mechanical strength is required such as in non-load bearing sites [5, 11]. On the other hand, several studies were carried out to investigate the possibilities of enhancing its mechanical properties [12, 13]. For instance, double-network hydrogels were examined by Yasuda and co-workers [14, 15], while Li et al. attempted to improve the mechanical properties by increasing the crosslinking density [16, 17] intended for applications such as cartilage regeneration.

In recent years, studies have demonstrated that the differentiation of stem cells is regulated by substrate stiffness [18–21], and stem cell differentiation would be promoted down particular lineages similar to 2D substrates with varying stiffness only if the 3D environments of the hydrogel mimic the stiffness of native tissue [18]. A study was carried out to investigate the possible link between the bulk matrix stiffness of the hydrogel and the differentiation of stem cells. It was found that the adipogenic and osteogenic differentiation of both human adipose-derived and mesenchymal stem cells plated onto hydrogels with different stiffness were regulated by the mechanical feedback offered by hydrogel deformations on planar matrices independent of substrate porosity and protein tethering [19]. Mooney and co-workers demonstrated the progression of wound healing could be controlled by adjusting the matrix mechanical properties such as a hydrogel deposited at the wound site when they examined the possibility of manipulating the behavior of dermal fibroblasts by tuning the stiffness of a model wound dressing biomaterial [20]. Similar observations were made in their earlier study in which they demonstrated a change in the commitment of mesenchymal stem cell populations in response to the rigidity of 3D environments. The study also revealed matrix stiffness regulated integrin binding in addition to reorganization of adhesion ligands on the nanoscale, both of which were traction-dependent and correlated with osteogenic commitment of mesenchymal stem cell populations [21].

In addition, observations from several studies suggested that stem cells continuously respond to changes in the mechanical properties of the matrix based on their capacity of recollecting mechanical cues from past physical environments [18, 22, 23]. In 2014, a study was attempted to explore the notion of whether stem cells remember past physical signals and if these can be utilized to dose cells mechanically [22]. It was carried out based on the findings from an earlier study by Gilbert et al. that demonstrated the elasticity or stiffness of the substrate is a potent regulator in determining the fate of stem cells in culture [23]. The in vivo examination showed stem cells isolated from muscles cultured on soft hydrogel substrates that resembled the stiffness of muscle self-renewed in vitro and contributed extensively to muscle regeneration when subsequently transplanted into mice [23]. Likewise, Yang et al. noticed the length of exposure of mesenchymal stem cells on stiff polystyrene favored their differentiation toward osteogenesis during culture on soft hydrogels instead of adipogenesis originally [22].

1.1 Crosslinking Hydrogels

To avoid dissolution of the hydrophilic polymer chains in an aqueous environment, the presence of crosslinks is essential in a hydrogel. In general, hydrogels can be divided into two groups based on the nature of crosslinking used in their preparation [24].

1.1.1 Physically Crosslinked Hydrogels

Physically crosslinked hydrogels have gained popularity in recent years and the interest was driven by the fact that the necessity of utilizing crosslinking agents during its preparation is eliminated [24]. These crosslinking agents need to be removed from the hydrogels prior to their applications in the biomedical arena as they are often toxic compounds. In addition, it has been suggested that they can influence the integrity of the biological substance such as cells and proteins to be encapsulated within the hydrogel [24]. Physically crosslinked hydrogels display linkages that are naturally created as a result of entanglements in addition to other interactions available in the polymeric chain such as ionic interactions, crystallization (freeze-thawing), hydrogen bonding, and protein interactions, which makes them exist temporarily and reversible in nature [24, 25]. Furthermore, physically crosslinked hydrogels can also be obtained from amphiphilic block and graft copolymers, and according to the review by Hennink and van Nostrum, graft copolymers can be created using a water-soluble polymer backbone such as a polysaccharide to which hydrophobic units can be attached, or hydrophobic chains containing water-soluble grafts [24]. Warming a cool polymer solution, or conversely, cooling a hot polymer solution is a simple approach to synthesize physically crosslinked hydrogels and the latter can be used to produce gels from gelatin or agarose [7].

Physically crosslinked hydrogels often display reversible sol–gel transition in response to changes in external stimuli such as pH, temperature, through the application of stress, or addition of certain solutes which compete with the polymeric ligand for the affinity site on the protein [7, 26, 27]. In addition, ionically- or hydrophobically related domains as well as clusters of molecular entanglements contributed to the inhomogeneities found within physically crosslinked hydrogels. Transient network defects in the hydrogel can also be caused by free chain ends or chain loops [7].

1.1.2 Chemically Crosslinked Hydrogels

Hydrogels are referred to as "chemical" or "permanent" if they are covalently crosslinked, and it differs to physically crosslinked hydrogels in which the hydrogen bonds are replaced by more stable and stronger covalent bonds [24]. Hydrogels can be crosslinked chemically using compounds such as dialdehydes, glutaraldehyde, and formaldehyde [25]. Crosslinking of water-soluble polymers could also be used to create chemical crosslinked hydrogels. Another approach is to create a network using hydrophilic polymers converted from hydrophobic polymers together with crosslinking [7].

The capacity of a hydrogel to absorb water via its hydrated porous structure can be further enhanced through chemical crosslinking [25]. Because of their inhomogeneity similar to physical crosslinked hydrogels, chemical crosslinked hydrogels typically consist of clusters, which are area of high crosslink density and low water swelling, dispersed within areas of low crosslink density and high swelling [7]. A hydrogel with soft polymeric structures is the product of low crosslink density leading to a greater swelling ability, and similarly, high crosslink density produces a hydrogel with a structure that is more compact and less likely to swell [25]. Additionally, the formation of water-filled "macropores" or "voids" and the separation of phases can take place under certain scenarios based on several factors such as the temperature, composition of solvent, and concentrations of solids throughout gel formation [7].

In terms of the mechanical properties, free chains end within chemical crosslinked gels are simply defined as gel network "defects" and have no significant contribution to the elasticity of the network. Likewise, the elasticity of the network is not influenced by other network defects such as chain "loops" and entanglements [7].

Various options are available to obtain chemically crosslinked hydrogels. For instance, low molecular weight monomers can be used as a starting material and the gel is created via radial polymerization in the presence of crosslinking agents. This approach has been used to obtain hydrogels such as dextran from various water-soluble natural, synthetic, and semi-synthetic polymers [24]. Another approach is via chemical reactions of complementary groups, for instance, glutaraldehyde can be used to crosslink water-soluble polymers with hydroxyl groups such as poly(vinyl alcohol) (PVA) [28–30]. However, the obtained hydrogel is not ideal for clinical applications as studies by Ballantyne and co-workers have demonstrated the correlation between the applications of glutaraldehyde and oncogenicity and physiological toxicity using animal models [31–33]. Consequently, this has led researchers to identify a less toxic alternative that can be used as a crosslinking agent for PVA hydrogels and studies were carried out to examine the possible use of dialdehyde cellulose as a replacement for glutaraldehyde in areas such as drug delivery and wound dressing [34, 35]. It should also be mentioned that amine-containing polymers could be crosslinked under mild conditions using the same reagents [24].

Bis or higher functional crosslinking agents can also be utilized to convert water-soluble polymers into hydrogels through addition reactions whereby a reaction occurs between the agent and the functional groups of water-soluble polymers [24]. The quantity of the crosslinking agents used along with the concentration of the dissolved polymer influences the properties of the gel network. Since water can also react with the crosslinking agent, it is preferential to conduct the crosslinking process in organic solvents. More importantly, the hydrogels must be extracted at great lengths to remove traces of unreacted agents, as they are extremely toxic [24].

Condensation reactions are another crosslinking approach that can be used for the preparation of hydrogels. It has been stated that condensation reactions between hydroxyl and amino groups with carboxylic acids or derivatives are often applied in the synthesis of polymers to yield polyesters or polyamides, respectively [24]. N-(3-dimethylaminopropyl)-N'-ethylcarbodiimide hydrochloride (EDC) is often used to crosslink water-soluble polymers with amide bonds. It has been discovered that EDS is able to alter side-groups on proteins to render them reactive with other side-groups and to mediate the formation of ester bond between hydroxyl and carboxyl groups of natural polymers such as hyaluronic acid and has found widespread use in crosslinking gelatin, collagen, as well as hyaluronic acid [36]. This approach was used in several recent studies in an attempt to prepare hydrogels that could be used in areas such as non-adherent wound dressing and tissue regeneration membrane [37, 38].

1.1.2.1 High Energy Radiation

As discussed earlier, it is of vital importance that hydrogels must possess adequate biocompatibility and the products formed as a consequence of their degradation have low toxicity if they were to be implanted into the human body or used as a drug delivery system. This implies that the formed degradation compounds must be able to be metabolized into harmless products or excreted through renal filtration process [24]. In addition to the various chemical methods discussed above in which the use of toxic crosslinking agents is necessary resulting in the possibility that residual agents can be detected in the hydrogels, an interesting and different approach is offered by ionizing high energy radiation such as electron beam, gamma rays, and UV radiation [39–48]. Water-soluble polymers can be crosslinked using high-energy irradiation without additional vinyl groups. Radicals can be created on the polymer chain through the hemolytic clipping of C–H bonds throughout the course of gamma or electron beam irradiation of polymers from aqueous solutions. In addition, the process can be carried out under physiological pH and at room temperature in water. Most importantly, the use of toxic crosslinking agents is avoided. The radiation dose along with the concentration of the polymer will govern the permeability and swelling properties of the formed hydrogel, and in general, increasing the dose of radiation and the concentration of the polymer will increase the crosslink density [24].

Besides gamma rays, ultraviolet (UV) or visible light can also interact with photoinitiators to form free radicals which can initiate polymerization to create crosslinked hydrogels [48]. In comparison with conventional polymerization approaches, photopolymerization offers several advantages such as the rate of curing is relatively fast at physiological or room temperature, the amount of heat produced is kept to a minimum, and the polymerization process can be controlled temporally and spatially [48, 49]. Photocrosslinkable hydrogels have been the subject of numerous in vitro and in vivo investigations as tissue engineering scaffolds to induce cartilage formation as well as other applications in regenerative medicine [50–62].

1.1.2.2 Enzymes

Enzymatic-mediated hydrogels are covalently bonded through reactions which are catalyzed by certain enzymes based on the polymer used [63]. The use of enzymes to create covalently crosslinked hydrogels can resolve a number of drawbacks and limitations created by some of the more commonly used crosslinking approaches such as the insufficient mechanical stability and strength associated with physically crosslinked hydrogels or the potential of cytotoxicity in photoinitiator-based hydrogels [64]. Most of the enzymes that take part in the crosslinking process are similar to those used during catalyzing reactions that occur naturally in the human body. Furthermore, enzymatic reactions are catalyzed by most enzymes in aqueous solutions, at physiological temperatures, and at neutral pH, suggesting that they can be used to develop in situ forming hydrogels which are extremely desirable for biomedical applications [64, 65]. Because of the substrate specificity of the enzyme, which is also one of the best advantages with enzymatic reactions, undesirable toxicity or side reactions which can arise from reactions that utilize organic solvents or photoinitiators are averted. The mildness of the enzymatic reactions at normal physiological conditions is another key advantage, indicating that this approach can be employed to crosslink natural polymers that are incapable of withstanding the harsh conditions faced during chemical crosslinking. In addition, enzymatic reaction also reduces the likelihood of bioactivity being reduced [64]. However, despite the advantages offered by enzymatic crosslinking of hydrogels for biomedical applications, numerous limitations need to be addressed and they are all based on reaction conditions. Factors such as temperature and pH have been postulated to influence the catalytic activity of enzymes in a significant manner rendering the crosslinking process difficult to use [66].

Numerous enzymes have been successfully used to induce hydrogel crosslinking, and one of the primary enzyme-mediated crosslinking approaches is based on the oxidation reaction catalyzed by peroxidase such as horseradish peroxidase or tyrosinase [66–68]. Horseradish peroxidase is a widely used enzyme to crosslink phenol residues and typically, they are combined with hydrogen peroxide in which the oxygen from the peroxide is extracted resulting in a change in the oxidation state of its heme group. Hydrogels are formed through the crosslinking of the aromatic ring by C–O and C–C coupling via the actions of hydrogen peroxide combined with horseradish peroxidase in the presence of phenolic hydroxyl groups [68]. Recently, crosslinking of regenerated silk with tyramine-substituted silk or gelatin was attempted to enhance mechanical properties and gelation kinetics of silk hydrogel through an increase in the number of phenol groups available for enzymatic crosslinking [69]. The study also claimed they were able to adjust the bioactivity of the silk hydrogels by changing the cyclo (RGCyK) or gelatin content which enabled the alteration of the cell microenvironment for specific cell delivery or scaffold applications. In addition, the hydrogel disks were also implanted subcutaneously in mice and no adverse effects from the implants were recorded over three days. Later, horseradish peroxidase-crosslinked silk fibroin hydrogels using calcium peroxide as oxidizer were attempted and the in vitro investigation revealed the hydrogel was able to promote SaOs-2 tumor cell's death after 10 days of culturing, upon complete β-sheet conformation transition. The angiogenic potential was also examined in an in ovo chick chorioallantoic membrane (CAM) assay, which showed a high number of converging blood vessels but no endothelial cells infiltration. In vivo examination using CD1 and nude NCD1 mice revealed the hydrogel was stable up to 6 weeks post-implantation [70].

A study by Li et al. postulated that the combination of a 3D-printed macroporous hydrogel scaffold created through horseradish peroxidase-mediated crosslinking of silk fibroin and tyramine-substituted gelatin would offer a solution for the repair and regeneration of cartilage tissues. In vivo observations using rabbit models showed the combination of hydrogel and human adipose-derived cell aggregates promoted articular cartilage regeneration 12- and 16-weeks post-implantation [71].

The repair effect of an injectable multifunctional carboxymethyl chitosan/ hyaluronic acid-dopamine hydrogel crosslinked by horseradish peroxidase and hydrogen peroxide on full-thickness skin injury model in mice was attempted, and observations on wound healing, collagen deposition, immunohistochemistry and immunofluorescence suggested that the hydrogel could potentially promote angiogenesis and cell proliferation at the injured site [72]. Similarly, an injectable hyaluronic acid hydrogel dual-enzymatically crosslinked in situ by galactose oxidase and horseradish peroxidase was evaluated for their therapeutic effect on mice with traumatic brain injury. The authors stated that the tyramine-modified hydrogel loaded with bone mesenchymal stem cells and nerve growth factor could potentially facilitate the survival and proliferation of endogenous neural cells possibly by neurotrophic factors release and neuroinflammation regulation and subsequently enhance the neurological function recovery and accelerate the repair process [73].

In addition to horseradish peroxidase, hematin has also been suggested as an alternative enzyme. This is based on the notion that horseradish peroxidase is a plant-derived protein, which may provoke an immunological response, and consequently, the safety of horseradish peroxidase must be verified prior to any clinical implantation of horseradish peroxidase-crosslinked hydrogels into human subjects [65]. Hematin is a porphyrin-containing iron with an iron (III) compound structure bearing similarity to the pseudo-iron protoporphyrin IX moiety found in horseradish peroxidase. More importantly, it can be used for the oxidative crosslinking of phenolic compounds in the presence of hydrogen peroxide [68, 74]. The study by Sakai et al. demonstrated that no adverse effects were noticed when hematin was used in comparison with horseradish peroxidase based on histological observations of cutaneous tissue surrounding in situ formed gels [74]. In 2022, a study was attempted to synthesize hematin-gelatin-tyramine as a precursor biocompatible polymer that could be exploited to prepare an injectable hydrogel in the absence of horseradish peroxidase [75]. Even though the hydrogel demonstrated adequate biodegradability, further investigations on its biocompatibility are required.

Tyrosinase is a binuclear copper-containing ubiquitous oxidative enzyme that can convert low molecular weight phenols or accessible tyrosyl residues of proteins into quinones [76]. Due to their mild reaction conditions, tyrosinase is increasingly being investigated for the crosslinking of hydrogels and in situ cell encapsulation [67, 77]. Since it can oxidize most substrates with monophenol residues, tyrosinase possesses the advantage that various materials can be used and the tyrosinase-mediated crosslinked hydrogel display adhesiveness to materials with phenol, N=NH, and –SH groups such as those discovered in skin tissues. Furthermore, catechol and quinone groups formed by oxidation are reactive and they may possibly play a role in hydrogel formation using different categories of reactions as well as in the adhesive reactions even though they do not contribute to the crosslinking process [66].

Observation from a previous study has shown the capacity of tyrosinase in initiating gel formation with blends of gelatin and chitosan through the reactions of tyrosyl residues of gelatin and tyrosinase [78]. The study also revealed tyrosinase can oxidize gelatin based on dissolved oxygen and spectrophotometric investigations and the authors estimated around 20% of the gelatin chains would undergo reaction. Above all, the gel formed from tyrosinase had different thermal behavior and were broken by the chitosan-hydrolyzing enzyme chitosanase in comparison with gelatin gel [78].

In 2018, a study was attempted to synthesize a gelatin-hyaluronic acid tissue adhesive hydrogel crosslinked using a recombinant tyrosinase from *Streptomyces avermitilis* that can be sprayed using commercial airbrush nozzle or injected with a medical syringe [79]. According to the authors, tyrosinase from *Streptomyces avermitilis* has a flat, wide, and shallow active site that broadens the substrate specificity and displays superior reactivity on long-chain polymers, which accelerates polymer crosslinking when compared to other tyrosinase. The study also claimed the amount of time needed to reach complete gelation was less than 50 s, and the physical properties and adhesive strength of the hydrogel were noticeably improved with the native tissue samples. In addition, the hydrogel displayed negligible immune reaction during in vivo analysis. The concept of injectability was further explored by the same group when they fabricated inflammation-modulating composite hydrogels that could provide a therapeutic option for the treatment of osteoarthritis [80]. Epigallocatechin-3-gallate (EGCG), which is found in green tea and is one of the polyphenols that can scavenge radical oxygen species and prevent inflammation-induced oxidative damage, was incorporated into the hydrogel and in vitro observations revealed it might offer protection to chondrocytes against the pro-inflammatory factor IL-1β and could result in chondrogenic regeneration. In addition, the injection of EGCG-loaded composite hydrogel to surgically induced osteoarthritic knee joints demonstrated its retention capability in the joint and the anti-inflammatory EGCG component could be maintained within the surface of articular cartilage for an extended period due to the adhesive properties of the hydrogel.

A different approach was utilized by Öztürk et al. for the fabrication of tissue adhesive hydrogel [81]. Their study centered on the sequential modification of alginate with first sulfate moieties to mimic the high glycosaminoglycan content of native cartilage and then tyramine moieties to allow in situ enzymatic crosslinking with tyrosinase under physiological conditions. Based on their findings, the authors stated the hydrogels displayed strong adhesion to native cartilage tissue, supported the viability of encapsulated bovine chondrocytes, and induced a strong increase in the expression of chondrogenic genes such as collagen 2, aggrecan, and Sox9. Chondrocytes in the hydrogels showed potent deposition of cartilage matrix components collagen 2 and aggrecan after 3 weeks of culture. The hydrogels with encapsulated human chondrocytes showed in vivo stability as well as cartilage matrix deposition upon subcutaneous implantation into mice for 4 weeks.

Enzymatic crosslinked hydrogel nanofilms that could potentially be used for cell-based therapies were recently attempted, and according to the study, such nanofilms could protect cells from high shear stress and decrease immune response through the interference of cell–cell interaction [82]. Hydrogel nanofilm was fabricated by monophenol-modified glycol chitosan and hyaluronic acid that crosslink each other to form a nanothin hydrogel film on the cell surface via tyrosinase-mediated reactions. Furthermore, hydrogel nanofilm formation was conducted on mouse β cell spheroids for the islet transplantation application. According to the authors, the hydrogel nanofilm was able to act as a physical barrier against external pressure in addition to reducing cell–cell interaction with natural killer cell. The transplantation of the caged mouse β cell spheroids into the type 1 diabetes mouse model showed it could regulate its blood glucose level.

Apart from the utilization of tyrosinase to crosslink hydrogels, transglutaminase, a naturally occurring protein enzyme, has also been investigated. This enzyme is calcium-dependent, is ubiquitous in nature, and can be discovered in many species of the animal and plant kingdoms including humans [83]. Distributed extensively in various tissues, body fluids, and organs, transglutaminase catalyzes the formation of a covalent bond between the γ-carboxamide group of protein or peptide-bound glutamine and a free amine group. Aside from the formation of these bonds which display high resistance to proteolytic degradation, insoluble protein biopolymers that are extensively crosslinked are also created by transglutaminase which are indispensable for the organism to generate barriers and stable structures [83]. However, the broader applications of transglutaminase have been hampered due to the extremely high cost related to the acquisition of the enzyme from animal origin, and subsequently, the quest to acquire an enzyme of microbial origin has begun [84]. In 1989, microbial transglutaminase was first isolated from the *Streptoverticillium* sp. strain and researchers discovered that transglutaminase isolated from *Streptoverticillium mobaraense* did not require calcium ions for its activities [85]. In comparison with transglutaminases isolated from animal tissues, transglutaminases of microbiological origin have low molecular weight. A pH of 5.5 and a temperature of 40 °C have been suggested to be the most ideal for the catalytic activity of transglutaminase [84, 86]. However, this is not the case for transglutaminase isolated from *Streptomyces* sp., which functions most effectively at a higher temperature of 45 °C and is not stable at 50 °C (since it loses 50% of its activity when heated for 30 min) and is very susceptible to heat in the presence of ethanol [84].

In 2003, the study by Chen et al. compared the differences between microbial transglutaminase and mushroom tyrosinase and their capacity to catalyze the formation of gels from solutions of gelatin and chitosan [76]. In contrast to tyrosinase-catalyzed reactions, they discovered strong and permanent gels could be created from gelatin solutions without the need for chitosan and at a quicker rate if transglutaminase was used. However, the availability of chitosan throughout the transglutaminase-catalyzed reaction will lead to a stronger gel. The authors also observed that the transglutaminase-catalyzed gels lost the ability to undergo thermally reversible transitions characteristic of gelatin, which is consistent with transglutaminase's ability to covalently crosslink proteins. In contrast, cooling below the gel-point of gelatin strengthened tyrosinase-catalyzed gel and the authors implied that the reaction has no effect on the ability of the gelatin to undergo a collagen-like coil-to-helix transition. Tyrosinase-catalyzed gels were transient as their Young's modulus peaked at about 5 h after which the gels broke spontaneously over the course of 2 days.

Enzymatic crosslinking reaction has the potential of creating stable gelatin hydrogels at physiological temperatures that are both physiological robust and biocompatible. Subsequently, several studies have been carried out to gain further insights into the biodegradation of microbial transglutaminase-crosslinked gelatin hydrogels and the in vitro metabolic activities of human cells such as embryonic kidney cells, adipose-derived stem cells, umbilical vein endothelial cells, and dermal fibroblasts encapsulated within the hydrogel [87–91]. This was motivated by the fact that enzymatic crosslinking reaction is milder in comparison with chemical and physical approaches which could result in cell death as well as avoiding concerns such as the release of residual quantities of chemical crosslinking agents like glutaraldehyde during degradation of tissue engineering scaffolds in vivo [87]. Moreover, findings from these studies could then be translated into the development of three-dimensional cellular scaffolds to deliver regenerative cells in a controlled manner [87]. Yung et al. noticed the hydrogel displayed no mass loss after submersion into a saline solution at 37 °C, suggesting that it is thermally stable. Furthermore, the degradation rates were found to be tunable with gelatin content, which could potentially be useful for time-released regenerative cell delivery or long-time cell encapsulation [87].

Similar observations were made in another study in which the authors noticed the gelatin concentration in the hydrogel not only affected the degradation rate but also had an influence on cell growth morphology and cell proliferation [92]. They further commented that low concentration gelatin in the hydrogel resulted in noticeable cell loss during their cell proliferation experiments. To simulate the in vivo transplant environment prior to implantation subcutaneously into an adult SD rat model, a 3D cell migration model was designed and observations showed that cell migration activities began two days after in vitro implantation and low concentration of gelatin in the hydrogel favored cell migration and the average migration distance of adipose-tissue-derived stromal cells encapsulated in the hydrogel decreased as the gel concentration increased resulting in a denser hydrogel. According to the authors, this implies that encapsulated cells were able to be released from the hydrogel and could invade surrounding hydrogels. In vivo implantation revealed the hydrogel was not cytotoxic and was able to maintain the gel state in the body for more than two weeks.

Accordingly, several investigations were carried out using in vivo animal models to examine the therapeutic and regenerative capacities of cell-laden hydrogels. Observations based on rabbit knee joint models suggested that the implantation of a gelatin hydrogel enriched with articular cartilage extracellular matrix and human adipose-derived stem cells could repair a full-thickness osteochondral defect. At 8 weeks post-surgery, a smooth articular surface with transparent new hyaline-like tissue was recorded from a macroscopic point of view. The authors also stated that inflammatory responses were hardly observed and sound chondrocytes were aligned in the newly formed chondral layer based on histological analysis [93].

Lu et al. studied the wound healing potential of a gelatin hydrogel system with human adipose stem cells using a Wistar rat burn injury wound model [94]. In vivo burn wound healing experiments revealed cell spheroid encapsulated into hydrogel has a higher efficacy in accelerating wound contraction, suggesting tissues could regenerate faster. They also claimed the use of stem cell spheroids could significantly improve the wound healing process in comparison with cell suspension encapsulated in the hydrogel, and a hydrogel with stem cell spheroids was much more effective in angiogenesis given the fact that it possessed the possibility of cell–cell signaling for the promotion of vascular generation. Self-assembled human adipose-derived stem cell spheroids were also used in a recent study to investigate the potential a cell-laden hydrogel to treat diabetic periodontal wounds and craniofacial defects [95]. Oral mucosal wounds and calvarial osseous defects were created in diabetic rats. Observations from the study showed stem cell spheroids retained stemness and morphology in the hydrogel. Cell-laden hydrogels were found to accelerate wound closure, epithelization, and keratinization in mucosal wounds. Furthermore, osteogenic differentiation markers were evidently expressed and cell proliferation was promoted after 7 days and bone formation was noticeably stimulated after 28 days.

1.1.2.3 Free Radical Polymerization

Chemically crosslinked hydrogels can be prepared through the free radical polymerization of low molecular weight monomers in conjunction with crosslinking agent. This versatile approach is commonly used to develop polymer-based hydrogels from certain monomers like vinyl lactams, amides, and acrylates as these polymers encompassed ideal functional groups, or have been functionalized with radically polymerization groups [12, 96]. It is a category of chain-growth polymerization and plays a role in the synthesis of nanocomposite hydrogels. The polymer is formed by the sequential addition of free radical groups using this approach, which also serve as templates for structural build-up. Furthermore, this technique follows the typical free radical polymerization in the following stages: initiation, propagation, chain transfer, and termination [12]. Non-conventional radical polymerization techniques such as reversible addition-fragmentation chain transfer polymerization (RAFT) and atom transfer radical polymerization (ATRP) are also available [97].

Free radical polymerizations using the thiol-norbornene click reaction have gained increasing attention due to the many benefits offered by click reactions including high selectivity and orthogonal reactions [98, 99]. Light or photoinduced free radical polymerizations that utilize a photoinitiator represent a rapidly expanding "green" technology owing to its solvent-free formulations and the process is carried out under the influence of ultraviolet (UV) and visible lights in water [98, 100, 101].

Typically, the monomers are activated by an initiator during chain-growth polymerization which creates free radicals using light and heat as an energy source. The activated monomers will continue to trigger neighboring monomers, while at the same time the polymer chain continues to grow [97]. This polymerization technique is relatively fast and can produce high molecular weight polymers in short period, and hence, this is used when large amounts of hydrogel need to be prepared [12, 97]. Radical polymerization can be categorized into four approaches and they are bulk polymerization, solution polymerization, suspension polymerization, and emulsion polymerization. The radical initiators are dissolved in the liquid monomer in bulk polymerization, producing an optical transparent and glassy gel with no porosity [97]. Solution polymerization, on the other hand, will produce porous gels and requires a solution comprised of the monomers, crosslinking agents, and an initiator that needs to be activated using a redox initiator system, ultraviolet radiation, or thermal radiation. Monomers in suspension polymerization are insoluble within the dispersion medium and exhibit as unstable droplets. Consequently, the addition of a stabilizer to the medium is necessary to improve its stability. Moreover, with the aid of an initiator, polymerization will occur inside these droplets [97].

Initiators such as ammonium persulfate, ferrous ammonium sulfate, ceric ammonium nitrate, benzoyl peroxide, 2,2-azobisisobutyronitrile (AIBN), and potassium persulfate are utilized in free radical polymerization to initiate the reaction process. First described by Wichterle and Lím [102], poly(2-hydroxyethyl methacrylate) (pHEMA) is a well-known and often studied hydrogel in the biomedical arena for applications such as bone tissue scaffolds and contact lenses that can be obtained via free radical polymerization using crosslinking agents such as ethylene glycol dimethacrylate, melamine triacrylamide, and melamine trimethylacrylamide and an initiator such as AIBN [24, 103].

Synthetic monomers can be grafted over natural polymers using free radical mechanisms. Free radical grafting of reactive monomers is a vital chemical modification approach and it centers on the reaction between a polymer and a vinyl-containing monomer or a combination of monomers that will be able to form grafts onto the polymer backbone [104]. The ultimate intention of grafting is to improve the properties of the polymer such as biodegradability and mechanical characteristics [103]. Natural polymers such as starch, chitosan, and alginate have been grafted with synthetic monomers in the presence of a crosslinking agent or an initiator such as potassium persulfate [103, 105].

Free radical polymerizations are promising due to the rapid gelation times (from seconds to several minutes), and synthetic hydrogels formed using this approach offer the possibility of creating injectable scaffolds for chondrocyte encapsulation [98]. However, Bryant et al. commented that the polymerization reaction could be impacted by the in vivo and in vitro environments where cell encapsulations are carried out, and subsequently, the properties of the resultant hydrogel are also altered [98, 106]. Studies have revealed that an interaction occurs between free radicals and cells during encapsulation, and the impact on cells will be governed by the type of free radicals created during polymerization [98]. Furthermore, exposing cells to radicals could affect its functions afterward despite cells can survive the process of encapsulation into hydrogels formed from free radical polymerization. There is also a possibility that cells could interfere with the formation of hydrogel during encapsulation based on the theory that cells will react with free radicals. The study by Chu et al. hypothesized that during the process of cell encapsulation, the formation of hydrogel will be impacted by the presence of cells leading to a reduction in crosslinking and ultimately a decrease in the mechanical properties either via cellular interactions with the macromers leading to their depletion in solution prior to hydrogel formation, or cellular interactions with free radicals where cell membranes effectively function as chain transfer agents and terminates propagating radicals [98]. They also noticed the formation of poly(ethylene glycol) (PEG) hydrogels was altered by chondrocytes during encapsulation, causing a reduction in the bulk and local crosslink density. Freshly isolated chondrocytes were shown to interact with hydrogel precursors, in part through thiol-mediated events between dithiol crosslinkers and cell surface free thiols, depleting crosslinker concentration and leading to a reduction in the bulk hydrogel crosslink density. This effect was more pronounced with increasing cell density at the time of encapsulation.

Using in vivo models, several studies were carried out to examine the issues of biocompatibility and cytotoxicity [107–110]. Das et al. examined the possibility of synthesizing a terpolymeric covalently crosslinked hydrogel of hyaluronate that has been grafted with 2-hydroxyethyl to provide mechanical properties and elasticity and crosslinked using gelatin methacryloyl to alter cellular interactions as a biomaterial for areas such as cartilage tissue engineering [107]. Light-activated free radical polymerization was used in another study to form methacrylated fibrinogen hydrogels in the presence of macromolecular crosslinking polymers synthesized from acrylated PEG [108]. Animal experiments from both studies suggested the polymerization reaction could synthesize hydrogels that are both biodegradable and biocompatible and were able to support tissue regeneration.

In addition to tissue regeneration, the potential of crosslinked hydrogels as drug delivery carriers was also investigated [109, 110]. Oral controlled release pH-sensitive hydrogels were prepared using a polymeric blend of agarose, Pluronic acid, methacrylic acid, glutaraldehyde as crosslinker, and benzoyl peroxide as initiator [109]. Cyclophosphamide, an alkylating anticancer drug mainly used for non-Hodgkin's lymphoma treatment was loaded into the prepared hydrogel. The pH-dependent monomer was used to impart pH-dependent swelling and drug release behavior. Albino male rabbits were administered loaded hydrogel given at appropriate body weight dose through oral gavage. The authors claimed analyses from blood samples and photomicrographs of the vital organs showed the hydrogels were biocompatible and did not display any signs of ocular, skin, or oral toxicity. Monolith/hydrogel composites were attempted in another study through sequential photoinitiated free radical polymerization, including the polymerization of multimethacrylate substituted dodecamine organic molecular cage for fabrication of the monoliths and the subsequent post-modification by the sequential photoinitiated free radical polymerization of acrylated gelatin [110]. Mouse alkali burn-induced corneal neovascularization models were adopted to investigate the effect of composites loaded with triamcinolone acetonide on inhibiting corneal neovascularization. According to the authors, in vitro and in vivo biocompatibility tests showed that composites exhibited no obvious cytotoxicity to human corneal epithelium cells, the cornea, and the conjunctiva. Alkali-burned mice experiments and tandem mass tags-labeled quantitative proteomics revealed that drug-loaded composites displayed a positive effect on curing the corneal neovascularization by inhibiting the fibrinolytic system.

1.1.2.4 Click Chemistry

Due to its high reactivity, superb selectivity, and mild conditions, "click" chemistry, proposed in 2001 by Sharpless et al. [111], has appeared as an interesting method to prepare hydrogels with variable patterns and dimensions [112]. "Click" chemistry is a high-yield, highly selective, rapid, and spontaneous chemical reaction between two molecules under mild conditions and its utilization enables hydrogels to be crosslinked via several approaches with physical and chemical properties customizable to suit the intended application [113]. Furthermore, quick and modular synthesis of hydrogels with near-ideal network structures is made possible by the reliability and high efficacy of the "click" reactions. Numerous "click" reactions have also allowed the efficient incorporation of biomolecular building blocks such as peptide sequences either during or after the fabrication of hydrogels [114].

Bio-orthogonal is defined as the reaction that occurs in the presence of macromolecules such as proteins, living cells, and drugs. The distinctive bio-orthogonality of "click" reaction makes the developed hydrogel highly compatible with encapsulated bioactives [112, 115].

First reported for the synthesis of hydrogels using "click" chemistry, copper-based click chemistry is becoming a less preferred option for biomedical applications such as regenerative medicine and tissue engineering due to toxicity from copper ion and the generation of reactive oxygen species [113]. One of the widely utilized metal-catalyzed approaches for preparing crosslinked hydrogels is copper-catalyzed azide-alkyne cycloaddition; however, despite its advantages such as producing high quantity of products with short gelation time, a major issue is the toxicity associated with copper catalyst and the release of copper ions after the reaction even though there have been suggestions that ethylene diamine tetra-acetic acid (EDTA) could potentially be used to remove copper ions from the hydrogel [115]. Moreover, any residual copper interacting with enzymes and other biomolecules could render them inactive [114]. Hence, the cytotoxicity of copper ions restricts the utilization of this reaction for in vivo applications as eliminating the use of toxic catalysts is of paramount importance for biomedical applications. This implies copper-catalyzed conventional click reactions are limited to laboratory use only for the synthesis of hydrogels [116, 117].

Subsequently, copper-free "click" chemistry has emerged as a useful technique for synthesizing hydrogels intended to be used in regenerative medicine and tissue engineering and it offers numerous functionalities without the use of any toxic catalysts or produce any toxic end-product after gel formation [115]. The process first described in 2007 is based on the strain-promoted cycloaddition between cyclooctyne derivatives and azide-containing molecules [118]. The copper-free "click" chemistry was designed to eliminate the cytotoxic metal catalyst, harmful free radicals, and initiators needed by conventional click chemistry [117–119]. In addition, it permits artificial chemical reactions to transpire on the surfaces as well as on the inside of living cells both in vitro an in vivo without harmful side effects [116].

Numerous metal-free click reactions have been explored for the preparation of hydrogels. To address the issue of using copper as a catalyst and the associated biological toxicity, Bertozzi et al. introduced an alternative bio-orthogonal reaction modeled on a modified Huisgen 1,2-dipolar cycloaddition of azides and alkynes which forms triazole products [120, 121]. Their modifications include positioning the alkyne within a strained cyclooctyne ring and the addition of propargylic fluorine atoms, both of which significantly accelerated the reaction rate. They further commented that since the strain-promoted cycloaddition between cyclooctynes and azides occurs at a rate similar to metal-catalyzed transformation, they could be referred to as copper-free "click" chemistry, and such a reaction can be carried out under physiological conditions and without the need for a catalyst.

Benefits offered by the "click" chemistry such as the high fidelity and orthogonality of the reaction and its compatibility with physiological conditions have driven researchers to utilize strain-promoted azide-alkyne cycloaddition to conduct hydrogel crosslinking in situ. Nevertheless, the long-term biocompatibility and short-term cytocompatibility of the crosslinked hydrogel are just as important [122]. The preparation and in vivo performance of injectable hydrogels crosslinked in situ have been attempted in several investigations [123–125]. In 2013, a study was carried out to synthesize hydrogels that were administered into mice subcutaneously and intraperitoneally using a double-barreled syringe containing hyaluronic acids chemically modified with both azide and cyclooctyne groups. Residue of the hydrogel was eradicated 21 days after subcutaneous administration, while it was eliminated 7 days after intraperitoneal administration [123]. Later, a PEG-based injectable hydrogel was attempted and the authors stated that the hydrogel was formed in situ after subcutaneously injected into mice model. Based on in vivo observations, the authors noticed the implanted hydrogel caused a mild inflammatory response and the surrounding tissues fully recovered one-week post-injection [124]. Han et al. investigated the clinical potential of hyaluronic acid-based hydrogels as an injectable scaffold to regenerate cartilaginous tissue in vivo. Cultured articular chondrocytes isolated from the knee joints of rabbits were encapsulated within the hydrogels and were immediately injected into the subcutaneous dorsum of athymic mice. Histological analysis of implants retrieved 5 weeks post-implantation demonstrated the hydrogels were capable of regenerating cartilaginous tissue, as evidenced by chondrocytes in lacunae [125].

Thiol-ene reaction is another copper-free "click" chemistry that has attracted the attentions of biomedical researchers for the preparation of hydrogels, as there are a few characteristics that make it a predominantly versatile process for biological applications such as the reactions often advance swiftly under mild conditions and it is biocompatible with cells and other biological molecules [126, 127]. Moreover, it is a relatively simple procedure to introduce thiols and alkenes to macromolecules for the synthesis of hydrogels in comparison with other functional groups [127]. In general, thiol-ene "click" chemistry reaction occurs between thiol-containing compounds with the alkene groups of the reactants. Thiols are easily obtained from the amino acid cysteine and can perform "click" reactions with numerous functional groups. Thiol-ene reaction offers several advantages such as a nearly 100% reaction yield and high selectivity. In addition, the reaction can be carried out in water since it is unaffected by oxygen and water. Given that it is a light-guided reaction, the process has good spatial control [113, 115, 127]. Thioethers are produced when thiol-containing compounds undergo a reaction with alkenes under the influence of thermal or photoinitiators [113, 127]. It has also been suggested that the internal spacing of certain hydrogels can be altered through the regulation of place, rate, and time of the reaction by light [113].

Thiol-Michael type reactions and radially mediated thiol-ene reactions are the two different mechanisms that normally occur when thiol reacts with alkenes, with the reaction mechanism decided by the reactive functional groups and the conditions of the solution used [127]. Typically, the thiol-Michael type reaction occurs at room temperature with no interference from water protons, thus making the reaction flexible to the synthesis of hydrogels that take place normally in aqueous solutions. Above all, thiol-Michael reactions are ideal for the preparations of injectable hydrogels given that in most cases, the reactions generate no by-products and can take place without the use of catalysts, but if catalysts are required, it will only be in small quantities [127]. Hydrogel scaffolds prepared via thiol-Michael type reactions for tissue repairs were first reported by Hubbell and co-workers [128], and since then, a few in vivo studies have been attempted and observations suggest it could potentially be utilized for areas such as cartilage regeneration and soft tissue substitutes [129, 130].

On the other hand, an initiator is needed for radically-mediated thiol-ene chemistry despite the reaction can also be carried out under mild aqueous conditions. Radicals are created by the initiator that kick-starts the reaction between select alkene functional groups and thiol. Furthermore, the initiator used will govern if oxidation–reduction, photochemical, or thermal process will be employed for the generation or radicals [127, 131].

Thiol-ene hydrogels are prepared via covalent crosslinking of hydrophilic polymers containing alkenes (such as vinyl sulfone, maleimides, and norbornene) and reactive thiols. In general, sulfhydryls (thiol functional groups) are attached to a macromer through reactions with functional groups such as amines, aldehydes, or ketones that are commonly identified on synthetic and natural polymers. Additionally, reactive thiol function groups have been created through the modifications of ketone or aldehyde groups on macromolecules using hydrazide derivative such as acetamino-4-mercaptobutyric acid hydrazide [127].

Since metal catalysts are not required, hydrogels intended for regenerative medicine and tissue engineering are primarily prepared via thiol-ene and thiol-yne reactions due to its non-toxicity [115, 126]. Recently, injectable conductive hydrogels with tunable degradability were fabricated in a study to examine its potential as implantable bioelectrodes [132]. Two different hydrogel systems based on their degradability were fabricated using thiol-ene reactions with PEG-tetrathiol and thiol-functionalized reduced graphene oxide. Hydrolyzable PEG diacrylate was used to attain a degradable hydrogel, while non-degradable hydrogel was prepared using PEG-dimaleimide. The authors revealed the hydrogels displayed excellent cell and tissue compatibility. Furthermore, in vivo observations revealed the hydrolysable conductive hydrogels undergone complete degradation 3 days after administration, while hydrogels formed using stable PEG-dimaleimide were able to retain its shape for up to 7 days.

Even though a number of crosslinking approaches are available for PEG hydrogels, step growth photopolymerization based on the orthogonal reaction between norbornene and thiol has attracted the attentions of biomedical researchers for bone regeneration applications [133–136]. According to Hunckler et al., ester-linked thiol-ene PEG hydrogels will experience rapid in vivo degradation after transplantation into the subcutaneous or intraperitoneal space, making their use limited for biomedical applications without modification. Moreover, they postulated rapid in vivo degradation would lessen if an amide linkage were used instead of ester linkage. The hydrogel would then be able to maintain long-term in vivo stability for months, and such a replacement would not sacrifice the key properties associated with the ester-linked thiol-ene PEG hydrogels [133].

Furthermore, owing to their rapid rate of reaction and efficiency under mild aqueous conditions, hydrogels formed by thiol-ene "click" reactions are appealing for local controlled release of therapeutics as the chemistry permits in situ formation of gels with adjustable properties often responsive to environment signals [127]. Photoinitiated thiol-ene reactions have been investigated in several studies that attempted to develop hydrogels that can be used as a wound dressing while at the same time function as a drug delivery system [137–139].

The Diels–Alder cycloaddition, as the most useful reactions in material design, is another "click" reaction where the use of metal catalyst is not required. The reaction takes place between a conjugated diene and a substituted alkene (or dienophile) during which six π-electrons rearrange to create a cyclic, six-membered product [140–142]. Even in the absence of catalysts, the reaction could still proceed under mild conditions and no toxic products are generated. Due to their hydrophobicity, the presence of water causes the Diels–Alder reactions to accelerate rapidly [140–142]. Given that two covalent bonds are formed between interacting pair of diene and dienophile, the strength of the resultant hydrogel is typically high [142].

Characteristics such as regioselectivity, thermal reversibility, partial stereospecificity and stereoselectivity, and attributability of the reaction to normal or inverse electron-demand type are displayed by the Diels–Alder reactions [142]. Due to the matching of the highest occupied molecular orbital of diene with the lowest unoccupied molecular orbital of dienophiles, the normal electron-demand Diels–Alder reaction centers on the electron-poor withdrawing dienophiles and electron-rich donating dienes [142, 143]. One of the most common Diels–Alder reactions is between maleimide and furan where maleimide is used as the dienophile and the furan is utilized as the diene during the preparation of hydrogels [142]. In-situ forming injectable hydrogels based on Diels–Alder reaction has been investigated for potential use in drug delivery [144, 145] and tissue regeneration [146–148]. However, significant swelling of the gel due to the slow gelation time is one of the drawbacks when crosslinking hydrogels using normal Diels–Alder reaction [142].

The inverse electron-demand Diels–Alder or IEDDA reaction was conceived to address some of the drawbacks associated with the electron-demand Diels–Alder reaction [149]. According to the study by Blackman et al., IEDDA is said to be the fastest bio-orthogonal reaction to date [150]. Similar to the electron-demand Diels–Alder reaction, the reaction takes place between a diene and a dienophile but the similarities end there. In IEDDA, an electron-rich dienophile reacts with an electron-poor diene [149]. In addition, the quick reaction kinetics of IEDDA have been theorized to be the result of the low energy gap between the highest occupied molecular orbital of the dienophile and the lowest unoccupied molecular orbital of the diene [151]. As a consequence of the high chemoselectivity of the IEDDA reaction, very low concentration of reactants can be used for their conjugation to large biomolecules [151]. Nevertheless, the specificity of the compounds used and their lower availability in comparison with the normal electron-demand Diels–Alder reaction is one of the disadvantages of IEDDA [142].

Catalysts are not needed for activation in IEDDA reaction, and it can take place in water, biological media, and organic solvents [150]. Only minute amounts of nitrogen gas are produced by the reaction [142]. As a result of the reactivity of norbornene to free sulfhydryl groups, IEDDA "click" reactions involving norbornene moiety are exceptionally useful in the creation of modularly crosslinked hydrogels [152]. Only a few studies were attempted using IEDDA hydrogel for in vivo applications such as the controlled release of encapsulated drugs and all of them are based on tetrazine-norbornene interactions [152–154]. The non-hydrolytically degradable nature of tetrazine-norbornene crosslinked hydrogels is in part due to the irreversibility of the IEDDA "click" reaction, unless if degradation motifs are deliberately added to the hydrogel network [152, 154]. The long-term hydrolytic stability was investigated in a study in which IEDDA "click" hydrogel of norbornene modified poly(L-glutamic acid) and tetrazine functionalized four-arm PEG was injected subcutaneously into the back of rats and in vivo observations revealed thin fibrous capsules were formed around the hydrogel and it was able to maintain its stability for up to 4 weeks post-injection [154]. For that reason, Dimmitt et al. attempted to develop hydrolytically degradable PEG-based IEDDA "click" hydrogels with pre-engineered and highly tunable degradation kinetics. The hydrogels were able to be injected and were highly compatible in vivo [152].

In oxime "click" chemistry, a reaction takes place between an amino-oxy group and a ketone or an aldehyde and is ideal for the formation of hydrogels [155]. The reactions are fast and possess similar orthogonal functionalities that are demonstrated in cells and biomolecules [115, 155]. Typically, water is the byproduct of this reaction and the use of toxic catalyst, external temperature, or UV light is not required [115]. A unique feature of oximes is that the stability of the linkage can be fine-tuned by virtue of the general reversibility of this ligation [156]. Furthermore, the reaction partners are stable in comparison with thiols [155]. According to the study by Grover et al., this approach, which is similar to other "click" chemistries, should permit for covalent incorporation of signaling molecules into hydrogels, and in particular, as peptides and proteins can be altered with relative ease with oxoamide, ketones, or amino-oxy groups [155]. Hydrogels crosslinked using oxime reactions have been shown to possess slow rates of degradation when more stable oximes were used [157]. An injectable hyaluronan-oxime crosslinked hydrogel was attempted in a recent study, and according to the authors, the hydrogel was stable in vivo for up to 28 days, after which it began to degrade with around half of the hydrogel lost after 56 days [158].

1.2 Concluding Remarks

During the preparation of hydrogels, selecting the most appropriate crosslinking approach will significantly influence the biocompatibility and mechanical integrity of the hydrogel and ultimately its suitability for use in tissue reconstruction or drug delivery. Various options are available to obtain chemically crosslinked hydrogels. The use of enzymes to create covalently crosslinked hydrogels can resolve several drawbacks and limitations created by some of the more commonly used crosslinking approaches such as the potential of cytotoxicity in photoinitiator-based hydrogels. Most of the enzymes that take part in the crosslinking process are similar to those used during catalyzing reactions that occur naturally in the human body. "Click" chemistry has also been used for the preparation of hydrogels with variable patterns and dimensions. Copper-free "click" chemistry has emerged as a useful technique for synthesizing hydrogels that offers numerous functionalities without the use of any toxic catalysts or produces any toxic product after gel formation, which is ideal for regenerative medicine and tissue engineering.

References

1. Sperling LH (1977) Interpenetrating polymer networks and related materials. J Polym Sci Macromol Rev 12:141–180
2. Hoare TR, Kohane DS (2008) Hydrogels in drug delivery: progress and challenges. Polymers 49:1993–2007
3. Singhal R, Gupta K (2016) A review: tailor-made hydrogel structures (classifications and synthesis parameters). Polym Plast Technol Eng 55:54–70
4. Ahmed EM (2015) Hydrogel: preparation, characterization, and applications: a review. J Adv Res 6:105–121
5. Saravanan S, Vimalraj S, Lakshmanan G et al (2019) Chitosan-based biocomposite scaffolds and hydrogels for bone tissue regeneration. In: Choi AH, Ben-Nissan B (eds) Marine-derived biomaterials for tissue engineering applications. Springer series in biomaterials science and engineering, vol 14. Springer, Singapore, pp 413–442
6. Park H, Park K (1996) Hydrogels in bioapplications. In: Ottenbrite RM, Huang SJ, Park K (eds) Hydrogels and biodegradable polymers for bioapplications. ACS Publication, Washington, pp 2–10
7. Hoffman AS (2002) Hydrogels for biomedical applications. Adv Drug Deliv Rev 54:3–12
8. Vallés E, Durando D, Katime I et al (2000) Equilibrium swelling and mechanical properties of hydrogels of acrylamide and itaconic acid or its esters. Polym Bull 44:109–114
9. El-Sherbiny IM, Yacoub MH (2013) Hydrogel scaffolds for tissue engineering: progress and challenges. Glob Cardiol Sci Pract 2013:316–342
10. Slaughter BV, Khurshid SS, Fisher OZ et al (2009) Hydrogels in regenerative medicine. Adv Mater 21:3307–3329
11. Zhu J, Marchant RE (2011) Design properties of hydrogel tissue-engineering scaffolds. Expert Rev Med Devices 8:607–626
12. Mantha S, Pillai S, Khayambashi P et al (2019) Smart hydrogels in tissue engineering and regenerative medicine. Materials 12:3323. https://doi.org/10.3390/ma12203323
13. Li X, Sun Q, Li Q et al (2018) Functional hydrogels with tunable structures and properties for tissue engineering applications. Front Chem 6:499. https://doi.org/10.3389/fchem.2018.00499

14. Yasuda K, Kitamura N, Gong JP et al (2009) A novel double-network hydrogel induces spontaneous articular cartilage regeneration in vivo in a large osteochondral defect. Macromol Biosci 9:307–316

15. Fukui T, Kitamura N, Kurokawa T et al (2014) Intra-articular administration of hyaluronic acid increases the volume of the hyaline cartilage regenerated in a large osteochondral defect by implantation of a double-network gel. J Mater Sci Mater Med 25:1173–1182

16. Li X, Chen S, Li J et al (2016) 3D culture of chondrocytes in gelatin hydrogels with different stiffness. Polymers 8:269. https://doi.org/10.3390/polym8080269

17. Li X, Zhang J, Kawazoe N et al (2017) Fabrication of highly crosslinked gelatin hydrogel and its influence on chondrocyte proliferation and phenotype. Polymers 9:309. https://doi.org/10.3390/polym9080309

18. Walters NJ, Gentleman E (2015) Evolving insights in cell-matrix interactions: elucidating how non-soluble properties of the extracellular niche direct stem cell fate. Acta Biomater 11:3–16

19. Wen JH, Vincent LG, Fuhrmann A et al (2014) Interplay of matrix stiffness and protein tethering in stem cell differentiation. Nat Mater 13:979–987

20. Branco da Cunha C, Klumpers DD, Li WA et al (2014) Influence of the stiffness of three-dimensional alginate/collagen-I interpenetrating networks on fibroblast biology. Biomaterials 35:8927–8936

21. Huebsch N, Arany PR, Mao AS et al (2010) Harnessing traction-mediated manipulation of the cell/matrix interface to control stem-cell fate. Nat Mater 9:518–526

22. Yang C, Tibbitt MW, Basta L et al (2014) Mechanical memory and dosing influence stem cell fate. Nat Mater 13:645–652

23. Gilbert PM, Havenstrite KL, Magnusson KE et al (2010) Substrate elasticity regulates skeletal muscle stem cell self-renewal in culture. Science 329:1078–1081

24. Hennink WE, van Nostrum CF (2002) Novel crosslinking methods to design hydrogels. Adv Drug Deliv Rev 54:13–36

25. Kaith BS, Singh A, Sharma AK et al (2021) Hydrogels: synthesis, classification, properties and potential applications—a brief review. J Polym Environ 29:3827–3841

26. Huang J, Jiang X (2018) Injectable and degradable pH-responsive hydrogels via spontaneous amino-yne click reaction. ACS Appl Mater Interfaces 10:361–370

27. Rosiak JM, Yoshii F (1999) Hydrogels and their medical applications. Nucl Instrum Methods Phys Res B 151:56–64

28. Morandim-Giannetti AA, Rubio SR, Nogueira RF et al (2018) Characterization of PVA/glutaraldehyde hydrogels obtained using Central Composite Rotatable Design (CCRD). J Biomed Mater Res B Appl Biomater 106:1558–1566

29. Dai WS, Barbari TA (1999) Hydrogel membranes with mesh size asymmetry based on the gradient crosslinking of poly(vinyl alcohol). J Membr Sci 156:67–79

30. Peppas NA, Benner RE Jr (1980) Proposed method of intracopdal injection and gelation of poly (vinyl alcohol) solution in vocal cords: polymer considerations. Biomaterials 1:158–162

31. Van Miller JP, Hermansky SJ, Losco PE et al (2002) Chronic toxicity and oncogenicity study with glutaraldehyde dosed in the drinking water of Fischer 344 rats. Toxicology 175:177–189

32. Ballantyne B, Myers RC (2001) The acute toxicity and primary irritancy of glutaraldehyde solutions. Vet Hum Toxicol 43:193–202

33. Ballantyne B, Jordan SL (2001) Toxicological, medical and industrial hygiene aspects of glutaraldehyde with particular reference to its biocidal use in cold sterilization procedures. J Appl Toxicol 21:131–151

34. Muchová M, Münster L, Capáková Z et al (2020) Design of dialdehyde cellulose crosslinked poly(vinyl alcohol) hydrogels for transdermal drug delivery and wound dressings. Mater Sci Eng C Mater Biol Appl 116:111242. https://doi.org/10.1016/j.msec.2020.111242

35. Münster L, Capáková Z, Fišera M et al (2019) Biocompatible dialdehyde cellulose/poly(vinyl alcohol) hydrogels with tunable properties. Carbohydr Polym 218:333–342

36. Park SN, Park JC, Kim HO et al (2002) Characterization of porous collagen/hyaluronic acid scaffold modified by 1-ethyl-3-(3-dimethylaminopropyl)carbodiimide cross-linking. Biomaterials 23:1205–1212

37. Kumar A, Wang X, Nune KC et al (2017) Biodegradable hydrogel-based biomaterials with high absorbent properties for non-adherent wound dressing. Int Wound J 14:1076–1087
38. Chang KC, Chen WC, Chen CH et al (2021) Chemical cross-linking on gelatin-hyaluronan loaded with hinokitiol for the preparation of guided tissue regeneration hydrogel membranes with antibacterial and biocompatible properties. Mater Sci Eng C Mater Biol Appl 119:111576. https://doi.org/10.1016/j.msec.2020.111576
39. Bustamante-Torres M, Pino-Ramos VH, Romero-Fierro D et al (2021) Synthesis and antimicrobial properties of highly cross-linked pH-sensitive hydrogels through gamma radiation. Polymers 13:2223
40. Godoy-Alvarez FK, González-Torres M, Giraldo-Gomez DM et al (2021) Synthesis by gamma irradiation of hyaluronic acid-polyvinyl alcohol hydrogel for biomedical applications. Cell Mol Biol 67:58–63
41. Jeong JO, Park JS, Kim EJ et al (2019) Preparation of radiation cross-linked poly(acrylic acid) hydrogel containing metronidazole with enhanced antibacterial activity. Int J Mol Sci 21:187. https://doi.org/10.3390/ijms21010187
42. Mori H, Hara M (2016) Clusters of neural stem/progenitor cells cultured on a soft poly(vinyl alcohol) hydrogel crosslinked by gamma irradiation. J Biosci Bioeng 121:584–590
43. Almeida JF, Ferreira P, Alves P et al (2013) Synthesis of a dextran based thermo-sensitive drug delivery system by gamma irradiation. Int J Biol Macromol 61:150–155
44. Bayramgil NP (2012) Synthesis, characterization and drug release behavior of poly(1-vinyl 1,2,4-triazole) hydrogels prepared by gamma irradiation. Colloids Surf B Biointerfaces 97:182–189
45. Terao K, Nagasawa N, Nishida H et al (2003) Reagent-free crosslinking of aqueous gelatin: manufacture and characteristics of gelatin gels irradiated with gamma-ray and electron beam. J Biomater Sci Polym Ed 14:1197–1208
46. Giammona G, Pitarresi G, Cavallaro G et al (1999) New biodegradable hydrogels based on an acryloylated polyaspartamide cross-linked by gamma irradiation. J Biomater Sci Polym Ed 10:969–987
47. Allcock HR, Kwon S, Riding GH et al (1988) Hydrophilic polyphosphazenes as hydrogels: radiation cross-linking and hydrogel characteristics of poly[bis(methoxyethoxyethoxy)phosphazene]. Biomaterials 9:509–513
48. Nguyen KT, West JL (2002) Photopolymerizable hydrogels for tissue engineering applications. Biomaterials 23:4307–4314
49. Decker C (1987) UV-curing chemistry: past, present, and future. J Coat Technol 59:97–106
50. Qu Y, He S, Luo S et al (2023) Photocrosslinkable, injectable locust bean gum hydrogel induces chondrogenic differentiation of stem cells for cartilage regeneration. Adv Healthc Mater 12:e2203079
51. Ji X, Lei Z, Yuan M et al (2020) Cartilage repair mediated by thermosensitive photocrosslinkable TGFβ1-loaded GM-HPCH via immunomodulating macrophages, recruiting MSCs and promoting chondrogenesis. Theranostics 10:2872–2887
52. Ko CY, Ku KL, Yang SR et al (2016) In vitro and in vivo co-culture of chondrocytes and bone marrow stem cells in photocrosslinked PCL-PEG-PCL hydrogels enhances cartilage formation. J Tissue Eng Regen Med 10:E485–E496
53. Beck EC, Barragan M, Tadros MH et al (2016) Approaching the compressive modulus of articular cartilage with a decellularized cartilage-based hydrogel. Acta Biomater 38:94–105
54. Lin H, Cheng AW, Alexander PG et al (2014) Cartilage tissue engineering application of injectable gelatin hydrogel with in situ visible-light-activated gelation capability in both air and aqueous solution. Tissue Eng Part A 20:2402–2411
55. Levett PA, Melchels FP, Schrobback K et al (2014) Chondrocyte redifferentiation and construct mechanical property development in single-component photocrosslinkable hydrogels. J Biomed Mater Res A 102:2544–2553
56. Francisco AT, Hwang PY, Jeong CG et al (2014) Photocrosslinkable laminin-functionalized polyethylene glycol hydrogel for intervertebral disc regeneration. Acta Biomater 10:1102–1111

57. Shin H, Olsen BD, Khademhosseini A (2012) The mechanical properties and cytotoxicity of cell-laden double-network hydrogels based on photocrosslinkable gelatin and gellan gum biomacromolecules. Biomaterials 33:3143–3152
58. Jeon O, Powell C, Ahmed SM et al (2010) Biodegradable, photocrosslinked alginate hydrogels with independently tailorable physical properties and cell adhesivity. Tissue Eng Part A 16:2915–2925
59. Chou AI, Nicoll SB (2009) Characterization of photocrosslinked alginate hydrogels for nucleus pulposus cell encapsulation. J Biomed Mater Res A 91:187–194
60. Dadsetan M, Szatkowski JP, Yaszemski MJ et al (2007) Characterization of photo-crosslinked oligo[poly(ethylene glycol) fumarate] hydrogels for cartilage tissue engineering. Biomacromolecules 8:1702–1709
61. Nettles DL, Vail TP, Morgan MT et al (2004) Photocrosslinkable hyaluronan as a scaffold for articular cartilage repair. Ann Biomed Eng 32:391–397
62. Bryant SJ, Nuttelman CR, Anseth KS (1999) The effects of crosslinking density on cartilage formation in photocrosslinkable hydrogels. Biomed Sci Instrum 35:309–314
63. Naranjo-Alcazar R, Bendix S, Groth T et al (2023) Research progress in enzymatically crosslinked hydrogels as injectable systems for bioprinting and tissue engineering. Gels 9:230. https://doi.org/10.3390/gels9030230
64. Teixeira LS, Feijen J, van Blitterswijk CA et al (2012) Enzyme-catalyzed crosslinkable hydrogels: emerging strategies for tissue engineering. Biomaterials 33:1281–1290
65. Badali E, Hosseini M, Mohajer M et al (2021) Enzymatic crosslinked hydrogels for biomedical application. Polym Sci Ser A 63:S1–S22
66. Song W, Ko J, Choi YH et al (2021) Recent advancements in enzyme-mediated crosslinkable hydrogels: in vivo-mimicking strategies. APL Bioeng 5:021502. https://doi.org/10.1063/5.0037793
67. Liu HY, Greene T, Lin TY et al (2017) Enzyme-mediated stiffening hydrogels for probing activation of pancreatic stellate cells. Acta Biomater 48:258–269
68. Roberts JJ, Naudiyal P, Lim KS et al (2016) A comparative study of enzyme initiators for crosslinking phenol-functionalized hydrogels for cell encapsulation. Biomater Res 20:30. https://doi.org/10.1186/s40824-016-0077-z
69. Hasturk O, Jordan KE, Choi J et al (2020) Enzymatically crosslinked silk and silk-gelatin hydrogels with tunable gelation kinetics, mechanical properties and bioactivity for cell culture and encapsulation. Biomaterials 232:119720. https://doi.org/10.1016/j.biomaterials.2019.119720
70. Pierantoni L, Ribeiro VP, Costa L et al (2021) Horseradish peroxidase-crosslinked calcium-containing silk fibroin hydrogels as artificial matrices for bone cancer research. Macromol Biosci 21:e2000425
71. Li Q, Xu S, Feng Q et al (2021) 3D printed silk-gelatin hydrogel scaffold with different porous structure and cell seeding strategy for cartilage regeneration. Bioact Mater 6:3396–3410
72. Cui L, Li J, Guan S et al (2022) Injectable multifunctional CMC/HA-DA hydrogel for repairing skin injury. Mater Today Bio 14:100257. https://doi.org/10.1016/j.mtbio.2022.100257
73. Wang L, Zhang D, Ren Y et al (2021) Injectable hyaluronic acid hydrogel loaded with BMSC and NGF for traumatic brain injury treatment. Mater Today Bio 13:100201. https://doi.org/10.1016/j.mtbio.2021.100201
74. Sakai S, Moriyama K, Taguchi K et al (2010) Hematin is an alternative catalyst to horseradish peroxidase for in situ hydrogelation of polymers with phenolic hydroxyl groups in vivo. Biomacromolecules 11:2179–2183
75. Nguyen NH, Bui QA, Dang LH et al (2022) Preparation of injectable free-HRP-mediated and self-catalyzed gelatin-tyramine hydrogel for biomedical applications. Vietnam J Chem 60:606–614
76. Chen T, Embree HD, Brown EM et al (2003) Enzyme-catalyzed gel formation of gelatin and chitosan: potential for in situ applications. Biomaterials 24:2831–2841
77. Choi S, Ahn H, Kim SH (2022) Tyrosinase-mediated hydrogel crosslinking for tissue engineering. J Appl Polym Sci 139:e51887. https://doi.org/10.1002/app.51887

78. Chen T, Embree HD, Wu LQ et al (2022) In vitro protein-polysaccharide conjugation: tyrosinase-catalyzed conjugation of gelatin and chitosan. Biopolymers 64:292–302
79. Kim SH, Lee SH, Lee JE et al (2018) Tissue adhesive, rapid forming, and sprayable ECM hydrogel via recombinant tyrosinase crosslinking. Biomaterials 178:401–412
80. Jin Y, Koh RH, Kim SH et al (2020) Injectable anti-inflammatory hyaluronic acid hydrogel for osteoarthritic cartilage repair. Mater Sci Eng C Mater Biol Appl 115:111096. https://doi.org/10.1016/j.msec.2020.111096
81. Öztürk E, Stauber T, Levinson C et al (2020) Tyrosinase-crosslinked, tissue adhesive and biomimetic alginate sulfate hydrogels for cartilage repair. Biomed Mater 15:045019. https://doi.org/10.1088/1748-605X/ab8318
82. Kim M, Kim H, Lee YS et al (2021) Novel enzymatic cross-linking-based hydrogel nanofilm caging system on pancreatic β cell spheroid for long-term blood glucose regulation. Sci Adv 7:eabf7832. https://doi.org/10.1126/sciadv.abf7832
83. Martins IM, Matos M, Costa R et al (2014) Transglutaminases: recent achievements and new sources. Appl Microbiol Biotechnol 98:6957–6964
84. Kieliszek M, Misiewicz A (2014) Microbial transglutaminase and its application in the food industry. A review. Folia Microbiol 59:241–250
85. Ando H, Adachi M, Umeda K et al (1989) Purification and characteristics of a novel transglutaminase derived from microorganisms. Agric Biol Chem 53:2613–2617
86. Ho ML, Leu SZ, Hsieh JF et al (2000) Technical approach to simplify the purification method and characterization of microbial transglutaminase produced form *Streptoverticillium ladakanum*. J Food Sci 65:76–80
87. Yung CW, Wu LQ, Tullman JA et al (2007) Transglutaminase crosslinked gelatin as a tissue engineering scaffold. J Biomed Mater Res A 83:1039–1046
88. Chen PY, Yang KC, Wu CC et al (2014) Fabrication of large perfusable macroporous cell-laden hydrogel scaffolds using microbial transglutaminase. Acta Biomater 10:912–920
89. Alarake NZ, Frohberg P, Groth T et al (2017) Mechanical properties and biocompatibility of in situ enzymatically cross-linked gelatin hydrogels. Int J Artif Organs 40:159–168
90. Hou S, Lake R, Park S et al (2018) Injectable macroporous hydrogel formed by enzymatic cross-linking of gelatin microgels. ACS Appl Bio Mater 1:1430–1439
91. Tsai CC, Hong YJ, Lee RJ et al (2019) Enhancement of human adipose-derived stem cell spheroid differentiation in an in situ enzyme-crosslinked gelatin hydrogel. J Mater Chem B 7:1064–1075
92. Yang G, Xiao Z, Ren X et al (2016) Enzymatically crosslinked gelatin hydrogel promotes the proliferation of adipose tissue-derived stromal cells. PeerJ 4:e2497. https://doi.org/10.7717/peerj.2497
93. Tsai CC, Kuo SH, Lu TY et al (2020) Enzyme-cross-linked gelatin hydrogel enriched with an articular cartilage extracellular matrix and human adipose-derived stem cells for hyaline cartilage regeneration of rabbits. ACS Biomater Sci Eng 6:5110–5119
94. Lu TY, Yu KF, Kuo SH et al (2020) Enzyme-crosslinked gelatin hydrogel with adipose-derived stem cell spheroid facilitating wound repair in the murine burn model. Polymers 12:2997. https://doi.org/10.3390/polym12122997
95. Tu CC, Cheng NC, Yu J et al (2023) Adipose-derived stem cell spheroid-laden microbial transglutaminase cross-linked gelatin hydrogel for treating diabetic periodontal wounds and craniofacial defects. Stem Cell Res Ther 14:20. https://doi.org/10.1186/s13287-023-03238-2
96. Malpure PS, Patil SS, More YM et al (2018) A review on hydrogel. Am J PharmTech Res 8:42–60
97. Madduma-Bandarage USK, Madihally SV (2021) Synthetic hydrogels: synthesis, novel trends, and applications. J Appl Polym Sci 138:e50376. https://doi.org/10.1002/app.50376
98. Chu S, Maples MM, Bryant SJ (2020) Cell encapsulation spatially alters crosslink density of poly(ethylene glycol) hydrogels formed from free-radical polymerizations. Acta Biomater 109:37–50
99. Hoyle CE, Bowman CN (2010) Thiol-ene click chemistry. Angew Chem Int Ed Engl 49:1540–1573

100. Tomal W, Ortyl J (2020) Water-soluble photoinitiators in biomedical applications. Polymers 12:1073. https://doi.org/10.3390/polym12051073
101. Awwad N, Bui AT, Danilov EO et al (2020) Visible-light-initiated free-radical polymerization by homomolecular triplet-triplet annihilation. Chem 6:3071–3085
102. Wichterle O, Lím D (1960) Hydrophilic gels for biological use. Nature 185:117–118
103. Bashir S, Hina M, Iqbal J et al (2020) Fundamental concepts of hydrogels: synthesis, properties, and their applications. Polymers 12:2702. https://doi.org/10.3390/polym12112702
104. Hu GH, Flat JJ, Lambla M (1997) Free-radical grafting of monomers onto polymers by reactive extrusion: principles and applications. In: Al-Malaika S (ed) Reactive modifiers for polymers. Springer, Dordrecht, pp 1–83
105. Sorour M, El-Sayed M, El Moneem NA et al (2013) Free radical grafting kinetics of acrylamide onto a blend of starch/chitosan/alginate. Carbohydr Polym 98:460–464
106. Nicodemus GD, Bryant SJ (2008) Cell encapsulation in biodegradable hydrogels for tissue engineering applications. Tissue Eng Part B Rev 14:149–165
107. Das D, Cho H, Kim N et al (2019) A terpolymeric hydrogel of hyaluronate-hydroxyethyl acrylate-gelatin methacryloyl with tunable properties as biomaterial. Carbohydr Polym 207:628–639
108. Simaan-Yameen H, Bar-Am O, Saar G et al (2023) Methacrylated fibrinogen hydrogels for 3D cell culture and delivery. Acta Biomater 164:94–110
109. Aslam M, Barkat K, Malik NS et al (2022) pH sensitive pluronic acid/agarose-hydrogels as controlled drug delivery carriers: design, characterization and toxicity evaluation. Pharmaceutics 14:1218. https://doi.org/10.3390/pharmaceutics14061218
110. Huang C, Qi X, Chen H et al (2022) Monolith/hydrogel composites as triamcinolone acetonide carriers for curing corneal neovascularization in mice by inhibiting the fibrinolytic system. Drug Deliv 29:18–30
111. Kolb HC, Finn MG, Sharpless KB (2001) Click chemistry: diverse chemical function from a few good reactions. Angew Chem Int Ed Engl 40:2004–2021
112. Jiang Y, Chen J, Deng C et al (2014) Click hydrogels, microgels and nanogels: emerging platforms for drug delivery and tissue engineering. Biomaterials 35:4969–4985
113. Li X, Xiong Y (2022) Application of "click" chemistry in biomedical hydrogels. ACS Omega 7:36918–36928
114. Yigit S, Sanyal R, Sanyal A (2011) Fabrication and functionalization of hydrogels through "click" chemistry. Chem Asian J 6:2648–2659
115. Gopinathan J, Noh I (2018) Click chemistry-based injectable hydrogels and bioprinting inks for tissue engineering applications. Tissue Eng Regen Med 15:531–546
116. Yoon HY, Lee D, Lim DK et al (2022) Copper-free click chemistry: applications in drug delivery, cell tracking, and tissue engineering. Adv Mater 34:e2107192. https://doi.org/10.1002/adma.202107192
117. Barker K, Rastogi SK, Dominguez J et al (2016) Biodegradable DNA-enabled poly(ethylene glycol) hydrogels prepared by copper-free click chemistry. J Biomater Sci Polym Ed 27:22–39
118. Baskin JM, Prescher JA, Laughlin ST et al (2007) Copper-free click chemistry for dynamic in vivo imaging. Proc Natl Acad Sci U S A 104:16793–16797
119. Sletten EM, Bertozzi CR (2008) A hydrophilic azacyclooctyne for Cu-free click chemistry. Org Lett 10:3097–3099
120. Agard NJ, Prescher JA, Bertozzi CR (2004) A strain-promoted [3 + 2] azide-alkyne cycloaddition for covalent modification of biomolecules in living systems. J Am Chem Soc 126:15046–15047
121. Chang PV, Prescher JA, Sletten EM et al (2010) Copper-free click chemistry in living animals. Proc Natl Acad Sci U S A 107:1821–1826
122. Xu J, Filion TM, Prifti F et al (2011) Cytocompatible poly(ethylene glycol)-co-polycarbonate hydrogels cross-linked by copper-free, strain-promoted click chemistry. Chem Asian J 6:2730–2737
123. Takahashi A, Suzuki Y, Suhara T et al (2013) In situ cross-linkable hydrogel of hyaluronan produced via copper-free click chemistry. Biomacromolecules 14:3581–3588

124. Jiang H, Qin S, Dong H et al (2015) An injectable and fast-degradable poly(ethylene glycol) hydrogel fabricated via bioorthogonal strain-promoted azide-alkyne cycloaddition click chemistry. Soft Matter 11:6029–6036
125. Han SS, Yoon HY, Yhee JY et al (2018) In situ cross-linkable hyaluronic acid hydrogels using copper free click chemistry for cartilage tissue engineering. Polym Chem 9:20–27
126. Lowe AB (2010) Thiol-ene "click" reactions and recent applications in polymer and materials synthesis. Polym Chem 1:17–36
127. Kharkar PM, Rehmann MS, Skeens KM et al (2016) Thiol-ene click hydrogels for therapeutic delivery. ACS Biomater Sci Eng 2:165–179
128. Patterson J, Hubbell JA (2010) Enhanced proteolytic degradation of molecularly engineered PEG hydrogels in response to MMP-1 and MMP-2. Biomaterials 31:7836–7845
129. Chang J, Tao Y, Wang B et al (2015) An in situ-forming zwitterionic hydrogel as vitreous substitute. J Mater Chem B 3:1097–1105
130. Liu S, Pu Y, Yang R et al (2020) Boron-assisted dual-crosslinked poly (γ-glutamic acid) hydrogels with high toughness for cartilage regeneration. Int J Biol Macromol 153:158–168
131. Hoyle CE, Lowe AB, Bowman CN (2010) Thiol-click chemistry: a multifaceted toolbox for small molecule and polymer synthesis. Chem Soc Rev 39:1355–1387
132. Park J, Lee S, Lee M et al (2023) Injectable conductive hydrogels with tunable degradability as novel implantable bioelectrodes. Small 19:e2300250
133. Hunckler MD, Medina JD, Coronel MM et al (2019) Linkage groups within thiol-ene photoclickable PEG hydrogels control in vivo stability. Adv Healthc Mater 8:e1900371. https://doi.org/10.1002/adhm.201900371
134. Iglesias-Echevarria M, Durante L, Johnson R et al (2019) Coaxial PCL/PEG-thiol-ene microfiber with tunable physico-chemical properties for regenerative scaffolds. Biomater Sci 7:3640–3651
135. Wojda SJ, Marozas IA, Anseth KS et al (2020) Thiol-ene hydrogels for local delivery of PTH for bone regeneration in critical size defects. J Orthop Res 38:536–544
136. Lao W, Fan L, Zhang Q et al (2023) Click-based injectable bioactive PEG-hydrogels guide rapid craniomaxillofacial bone regeneration by the spatiotemporal delivery of rhBMP-2. J Mater Chem B 11:3136–3150
137. Goh M, Kim Y, Gwon K et al (2017) In situ formation of injectable and porous heparin-based hydrogel. Carbohydr Polym 174:990–998
138. Soiberman U, Kambhampati SP, Wu T et al (2017) Subconjunctival injectable dendrimer-dexamethasone gel for the treatment of corneal inflammation. Biomaterials 125:38–53
139. Zhang T, Guo L, Li R et al (2023) Ellagic acid-cyclodextrin inclusion complex-loaded thiol-ene hydrogel with antioxidant, antibacterial, and anti-inflammatory properties for wound healing. ACS Appl Mater Interfaces 15:4959–4972
140. Gregoritza M, Brandl FP (2015) The Diels–Alder reaction: a powerful tool for the design of drug delivery systems and biomaterials. Eur J Pharm Biopharm 97B:438–453
141. Diels O, Alder K (2010) Synthesen in der hydroaromatischen Reihe. Eur J Org Chem 460:98–122
142. Morozova SM (2023) Recent advances in hydrogels via Diels–Alder crosslinking: design and applications. Gels 9:102. https://doi.org/10.3390/gels9020102
143. Nicolaou KC, Snyder SA, Montagnon T et al (2002) The Diels–Alder reaction in total synthesis. Angew Chem Int Ed Engl 41:1668–1698
144. Ilochonwu BC, Mihajlovic M, Maas-Bakker RF et al (2022) Hyaluronic acid-PEG-based Diels–Alder in situ forming hydrogels for sustained intraocular delivery of bevacizumab. Biomacromolecules 23:2914–2929
145. Ilochonwu BC, van der Lugt SA, Annala A et al (2023) Thermo-responsive Diels–Alder stabilized hydrogels for ocular drug delivery of a corticosteroid and an anti-VEGF fab fragment. J Control Release 361:334–349
146. Bi B, Ma M, Lv S et al (2019) In-situ forming thermosensitive hydroxypropyl chitin-based hydrogel crosslinked by Diels–Alder reaction for three dimensional cell culture. Carbohydr Polym 212:368–377

147. Yang Y, Zhu Z, Gao R et al (2021) Controlled release of MSC-derived small extracellular vesicles by an injectable Diels–Alder crosslinked hyaluronic acid/PEG hydrogel for osteoarthritis improvement. Acta Biomater 128:163–174

148. Zhu Y, Sun Y, Rui B et al (2022) A photoannealed granular hydrogel facilitating hyaline cartilage regeneration via improving chondrogenic phenotype. ACS Appl Mater Interfaces 14:40674–40687

149. Handula M, Chen KT, Seimbille Y (2021) IEDDA: an attractive bioorthogonal reaction for biomedical applications. Molecules 26:4640. https://doi.org/10.3390/molecules26154640

150. Blackman ML, Royzen M, Fox JM (2008) Tetrazine ligation: fast bioconjugation based on inverse-electron-demand Diels–Alder reactivity. J Am Chem Soc 130:13518–13519

151. Oliveira BL, Guo Z, Bernardes GJL (2017) Inverse electron demand Diels–Alder reactions in chemical biology. Chem Soc Rev 46:4895–4950

152. Dimmitt NH, Arkenberg MR, de Lima Perini MM et al (2022) Hydrolytically degradable PEG-based inverse electron demand Diels–Alder click hydrogels. ACS Biomater Sci Eng 8:4262–4273

153. Czuban M, Srinivasan S, Yee NA et al (2018) Bio-orthogonal chemistry and reloadable biomaterial enable local activation of antibiotic prodrugs and enhance treatments against *Staphylococcus aureus* infections. ACS Cent Sci 4:1624–1632

154. Zhang Z, He C, Chen X (2020) Injectable click polypeptide hydrogels via tetrazine-norbornene chemistry for localized cisplatin release. Polymers 12:884. https://doi.org/10.3390/polym12040884

155. Grover GN, Lam J, Nguyen TH et al (2012) Biocompatible hydrogels by oxime click chemistry. Biomacromolecules 13:3013–3017

156. Kölmel DK, Kool ET (2017) Oximes and hydrazones in bioconjugation: mechanism and catalysis. Chem Rev 117:10358–10376

157. Li Y, Wang X, Han Y et al (2021) Click chemistry-based biopolymeric hydrogels for regenerative medicine. Biomed Mater 16:022003. https://doi.org/10.1088/1748-605X/abc0b3

158. Baker AEG, Cui H, Ballios BG et al (2021) Stable oxime-crosslinked hyaluronan-based hydrogel as a biomimetic vitreous substitute. Biomaterials 271:120750. https://doi.org/10.1016/j.biomaterials.2021.120750

Chapter 2
3D, 4D Printing, and Bioprinting of Hydrogels

2.1 3D Printing

Additive manufacturing, commonly referred to as three-dimensional (3D) printing, is the first of numerous useful printing methodologies currently available for tissue engineering and regenerative medicine. 3D printing of hydrogels, because of its capacity to create complex, intricate, and highly customizable tissue engineering scaffolds able to promote cell infiltration and support adhesion, has become an attractive option for biomedical researchers [1].

Simply put, the process of 3D printing begins with the construction of a 3D model that the computer uses to print a 3D representation from thin cross-sectional layers on top of one another. The technology has evolved, and additional processes have been created since the earliest version of 3D printing. The printing of complex scaffolds such as internal channels can be achieved with ease using this approach. More importantly, the incorporation of temperature-sensitive substances such as biological agents and pharmaceuticals into the scaffold can be realized as the printing process is carried out at room temperature [2].

It has been suggested that multi "color" printing is another favorable characteristic of 3D printing in tissue engineering, in which each color ink can be positioned on a precise location [2]. Moreover, this feature permits the possibility of depositing multiple extracellular matrices, organizing multiple cell types, and applying point-to-point control over bioactive agents simultaneously for biological tissue production. Numerous biological agents such as proteins, living cells, DNA plasmids, polysaccharides, and peptides have been printed using 3D printing [2].

It has been well established that tissue engineering scaffolds serve as templates for cell adhesion as well as recruiting cells to infiltrate deep into a defect site. They should also provide support during the regeneration process by creating spaces for tissue ingrowth and adequate mechanical strength. Even though improvements in the production of scaffolds using conventional fabrication techniques have been made, limitations still exist such as regulating the dimensions and shape of the scaffold

© The Author(s), under exclusive license to Springer Nature Singapore Pte Ltd. 2024

A. H. Choi and B. Ben-Nissan, *Hydrogel for Biomedical Applications*, Tissue Repair and Reconstruction, https://doi.org/10.1007/978-981-97-1730-9_2

produced in addition to the porosity and pore size. Considered design-dependent, the porosity, geometry, and size of the scaffold can be regulated precisely during fabrication using 3D printing in comparison with conventional approaches [3].

Despite the advantages associated with 3D printing, choosing appropriate biomaterials to serve as biomaterial inks for 3D printing has been labeled as a crucial yet limiting aspect when it comes to the design and application using this approach [4]. Hydrogels are one of the most promising and feasible categories of ink materials for the fabrication of 3D porous tissue engineering scaffolds [1, 4, 5]. Unfortunately, pure hydrogel when used as printing ink does not possess sufficient mechanical stability and can undergo degradation relatively easily, and subsequently, this led to the use of hydrogel composites as printing ink with enhanced biofunctionality and mechanical characteristics. The amalgamation of different properties and functions that cannot be obtained from any single hydrogel can be achieved using the composite approach [1].

The general term that has been used to describe all the methodologies currently available to fabricate objects through the delivery of materials in a progressive manner is solid freeform fabrication [6]. Solid freeform fabrication has enabled the design and production of patient-specific complex 3D structures [2]. Numerous additive manufacturing techniques have been utilized to produce hydrogels appropriate for tissue engineering.

2.1.1 Laser-Based Approaches

In essence, most laser-based systems are ideal for the processing of hydrogels with selective laser sintering being the only exception, and the system benefits from the photopolymerization route as the foundation to create crosslinked scaffolds [6]. These systems can be classified according to parameters such as the ways in which laser beams are delivered as well as well as the source of the laser used [1]. Laser-based systems such as stereolithography and two-photon polymerization work by the deposition of light energy in predefined patterns and in sequence so that only prepolymers that can be photocrosslinked can be used to print crosslinked hydrogels [4, 6]. Despite advantages such as the capacity stereolithography can build and the adequate surface finishes, post-curing is frequently needed as the produced hydrogel is often weak and lacks mechanical stability. In addition, post-processing is also needed by some systems to ensure unwanted materials such as support structures are removed [1].

Recently, a new stereolithographic approach known as two-photon polymerization has emerged and this micro- and nanofabrication technique can produce architecturally precise and predetermined hydrogel scaffolds with spatial complexity and high resolution (Fig. 2.1) [7, 8]. According to the study by Liska et al., femtosecond laser pulses of 800 nm are utilized by the two-photon polymerization process that is focused on the volume of a photopolymer [9]. At that laser wavelength, the photopolymer is transparent. Given the quadratic dependence of the two-photon

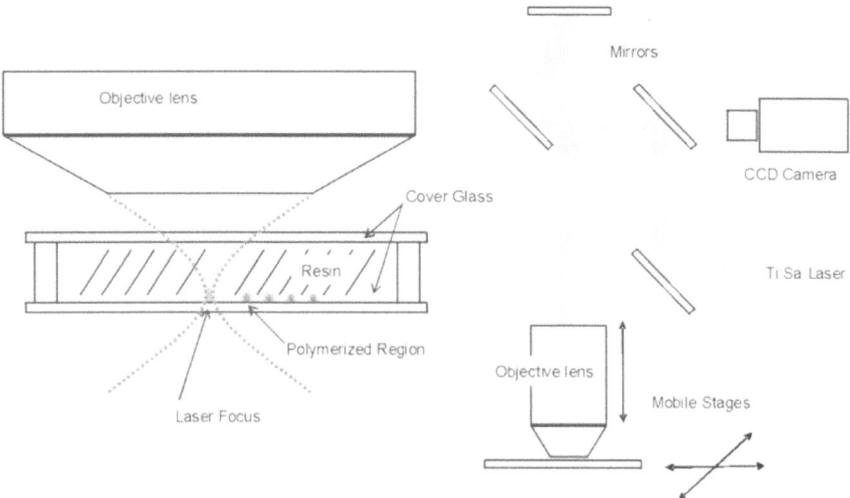

Fig. 2.1 Fabrication of 3D structures using two-photon polymerization. The laser is focused on the resin by an objective lens. The sample is mounted on a three-axis nanopositioning stage for controlled motion in all directions. Reprint with permission from [8]

absorption probability on the laser intensity, solidification is carried out in a highly localized volume. Once an ideal photoinitiator absorbs the two photons of 800 nm simultaneously, they serve as one 400 nm photon to initiate the polymerization process. Resolutions below the diffraction limit of the light used are possible with the two-photon polymerization [9].

It has also been suggested that moving the laser focus permits the synthesis of a direct "true" 3D object into the volume of the photosensitive material. It is also possible to create reproducible micron-sized objects with features less than 100 nm in size [6]. For that reason, two-photon polymerization is superior to all other solid freeform fabrication methodologies with respect to resolution and accuracy, and as such, hydrogel scaffolds produced by two-photon polymerization are gradually being investigated for tissue engineering applications [7, 10, 11]. Recently, a study was attempted to examine the biocompatibility of 3D scaffolds fabricated using a femtosecond laser with a wavelength of 780 nm to induce the two-photon polymerization effect [11]. Photosensitive gelatin methacrylate hydrogel solutions were constructed by combining photoinitiator, photosensitizer, and gelatin methacrylate to enhance the absorption efficiency in the near-infrared region.

Moreover, precise control over the 3D geometry of the scaffold is possible through the combination of two-photon polymerization and proper materials selection, which in turn supports the modeling and reproduction of cellular microenvironment [12]. The study by Ostendorf and Chichkov also postulated the cellular organization inside the scaffold can be regulated by the high resolution of the two-photon polymerization and subsequently, over cell interactions. They also suggested that the intensities of the near-infrared laser radiation used were found not to be harmful to cells, and this approach could potentially be utilized for the manipulation and encapsulation of cells [12]. The study by Weiss et al. also suggested the geometric parameters of scaffolds for 3D cell culture could be modified to cater for the special requirements of the target cell types [8].

Laser-induced forward transfer (LIFT) is a nozzle-free printing approach that has almost no restrictions on the viscosity of the ink or the size of particles to be printed. It is also capable of printing any functional material that can be formulated as an ink [13–15]. It has been suggested that LIFT technology is capable of transferring cells or bioinks onto a substrate in a precise manner, and this allows for the creation of complex 3D architectures with characteristics such as high printing precision and improved cell viability. Subsequently, it has found applications in bioprinting and the manufacture of biological structures and biomolecular microarrays. However, the high cost associated with this technology has limited its research and commercial application [16]. In addition, a number of attributes will regulate the laser printing process and these include the laser impact, the thickness of the hydrogel layer and its characteristics, and factors such as experimental configurations [17].

According to Serra and co-workers, LIFT system is composed of three components. A laser beam such as pulsed laser systems with pulses of several nanoseconds is focused by a lens that is situated on the donor substrate–layer interface. Most importantly, the donor substrate must be precoated with a layer of the ink or material that is to be printed and the donor substrate must be transparent at the laser wavelength. The focused laser beam causes part of the donor layer to be evicted and is deposited on the receiving substrate positioned in front of the donor layer. By moving the receiving and donor substrates and/or through the scanning of the laser beam, patterns can be printed by depositing pixels on the desired positions (Fig. 2.2) [13–15].

Given that the donor substrate must be transparent to the laser wavelength, glass can be used for visible and near-infrared lasers. For UV laser wavelength, quartz or fused silica is needed. A thin coating of the ink must be extended uniformly on the rear surface of the donor substrate and it must contain at least one component that absorbs the laser radiation. If not, either a thin absorbing layer must be placed between the

Fig. 2.2 Schematics of the main elements in a typical LIFT setup for laser printing. Reprint with permission from [15]. http://creativec ommons.org/licenses/by/4.0/

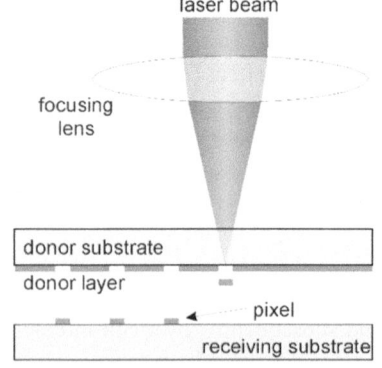

ink layer and the donor substrate or nonlinear absorption must be encouraged using femtosecond laser pulses. This thin absorbing intermediate layer, also referred to as the sacrificial layer, can be a thick polymeric layer that decomposes during transfer to reduce contamination or a nanometer-thick metallic film [13–15].

Bioprinting (which will be discussed later in greater details) offers the possibility of fabricating 3D living tissues and one of the most promising laser printing technologies is the LIFT of cell-laden bioinks [16, 18, 19]. This is performed through the precise layer-by-layer printing of biological materials such as living cells and cell-laden hydrogels. In 2004, Barron et al. demonstrated the ability of a modified LIFT approach that could accurately and rapidly print patterns of single cells in a non-contact manner. Human osteosarcoma cells were deposited into a biopolymer matrix. The printed cells demonstrated to be 100% viable after six days of incubation [20, 21]. They further commented that this modified LIFT approach was capable of rapidly depositing patterns of active biomolecules and living cells onto numerous material surfaces. In addition, this method could deliver small droplets of biomaterials (femtoliter to nanoliter) without the need for an orifice unlike manual spotting or inkjet techniques. Hence, the likelihood of clogging is reduced and the approach permits the deposition of diverse categories of biomaterials [20]. The LIFT approach can also be used for the fabrication of scaffold-free 3D cell systems via a layer-by-layer approach through the amalgamation of materials capable of forming stable gels and cell solutions. Moreover, it can also be applied to populate scaffolds with various cells and different cell densities in a precise manner. Above all, it was demonstrated that printed cells are not affected by the laser printing process and that differentiation of printed stem cells is not induced [22].

In 2019, a study was attempted to develop a new hybrid laser printing technology that could print 3D multiscale, multimaterial, complex hydrogel structures with microscale resolution [23]. According to the authors, this hybrid technology combined the key advantages of multiphoton polymerization and multiphoton ablation processes (superior design flexibility and high resolution) and the quick on-demand continuous fabrication of the continuous liquid interface production process. The authors also stated that this printing approach utilized sequential additive and subtractive modes of material fabrication that are normally thought of as mutually exclusive based on the variations in their material processing conditions. In addition, the new fabrication approach could print structures at any depth within the material in comparison with laser writing systems currently available that enforce stringent processing depth limits [23].

2.1.2 Extrusion or Nozzle-Based Approach

It has been said that extrusion or nozzle-based printing is the most widely used technique to construct hydrogel-based scaffolds, and in this approach, the viscous or melted hydrogels are extruded through a syringe or nozzle and a 3D object is built layer-by-layer via extrusion of materials in a sequential manner following a predesigned path created by computer modeling. More importantly, this technique requires sufficient adhesion between the printed layers and must be able to support its own weight before the next layer is deposited. Consequently, factors such as the gel setting mechanism, solidification temperature, and the rheological properties of the hydrogels are of paramount importance [1].

Extrusion or nozzle-based printing can be further categorized into two groups based on whether melting process is involved. Precise extrusion deposition and manufacturing, multiphase jet solidification, 3D fiber-deposition technique, and fused deposition modeling are some of the techniques based on melting processing [6, 24]. The melt process, from the aspect of bioactivity of the printed scaffold, is generally undesirable due to the use of elevated temperatures. As a result, a search began to replace melt process with dissolution of materials and examples of such systems include robocasting, pressure-assisted microsyringes, low-temperature deposition manufacturing, and multinozzle deposition manufacturing [24].

3D plotting, a low-temperature extrusion approach, is a much more versatile system compared to the fused deposition modeling process. Developed in 2000 at the Freiburg Materials Research Center, 3D plotting was envisaged to cater for the demands of desktop fabrication of scaffolds suitable for tissue engineering applications, and its ability to print hydrogel scaffolds with a well-defined internal pore structure is one of the most appealing characteristics of this methodology [25]. Another main feature is the 3D dispensing of liquids and pastes in liquid media with matched density, which compensates for gravity and subsequently, no support structures are required during the printing process (Fig. 2.3) [24]. In addition, a greater variety of both natural and synthetic materials including aqueous solutions and pastes can be used to fabricate scaffolds than the conventional 3D printing approaches such as those based on melting processing. Most importantly, this 3D plotting approach has been developed specifically for biofunctional processing, making it possible to integrate aqueous biosystems such as living cells into the scaffold fabrication process for the first time [26].

To fabricate scaffolds using this approach, viscous hydrogels are loaded initially into a syringe and injected into a liquid solution with a density matching that of the

Fig. 2.3 Schematic showing the principles of 3D-Bioplotting™ system. Reprint with permission from [6]

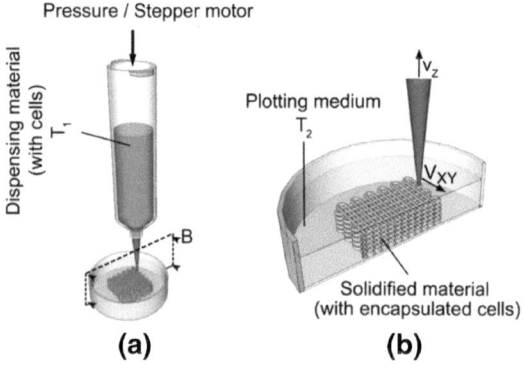

hydrogel through a microneedle that functions as the extrusion nozzle. Pneumatic nozzle utilizing filtered air pressure or volume-driven injection nozzle via a stepper motor can be used to generate liquid flow (Fig. 2.3). A continuous dispensing of microstrands or a discontinuous dispensing of microdots can be performed. The 3D construction of objects takes place in a laminar manner by the computer-controlled deposition of material on a surface. The fabrication platform is stationary while the dispensing heads move in three dimensions. Materials with low viscosity are particularly beneficial from this buoyancy compensation principle. 3D plotting can also be used to construct thermally sensitive biocomponents as well as cells since heating is not required. The thickness of the microstrand can be regulated by adjusting factors such as the rate of deposition in the planar direction, the magnitude of pressure applied, tip diameter, and the viscosity of the hydrogel. Curing reactions can be carried out via two-component dispensing using mixing nozzle or by plotting in a co-reactive medium [1, 6, 25]. However, surface treatment is needed for scaffolds printed using 3D plotting to make the surface more ideal for cell adhesion since constructs built using this technique mostly have smooth surfaces [25]. The in vitro performance of hydrogel tissue scaffolds fabricated using 3D plotting was compared to those manufactured using the freeze-drying approach. Observations suggested osteoblasts cell proliferation was better promoted in the 3D-plotted scaffolds, which were also stronger and less likely to degrade [27]. Recently, a study hypothesized an approach that could be utilized to enhance bone regeneration and address the issue of osteoinduction centered on porous gelatin-alginate scaffold fabricated by 3D plotting [28]. MicroRNA and a nanodelivery vector were loaded as releasable drugs inside the scaffolds and the combined influence of scaffold degradation and microRNA release was examined using an in vivo animal model. The authors postulate that such a scaffold could activate the osteoblastic differentiation of human bone mesenchymal stem cells and induce osteogenesis in vivo due to the presence of microRNAs.

2.1.3 Digital Light Processing

Digital light processing, a vat photopolymerization technology, can produce 3D hydrogels with finer accuracy and higher geometrical complexity and no substantial spatial resolution effect on the printing time. Localized solidification of a liquid formulation is caused by the projection of light, resulting in the fabrication of precise self-standing 3D structures layer-by-layer with fine spatiotemporal control [29].

First utilized by Lu et al., digital light processing (sometimes referred to as digital micromirror device-based stereolithography) is a quick and simple, layer-by-layer microstereolithography system that composed of an ultraviolet light source, a digital micromirror masking device, and a conventional computer projector that enables the fabrication of complex internal features in conjunction with precise spatial distribution of biological factors within a single scaffold (Fig. 2.4) [30]. They further commented that instead of writing the 3D microstructure point-by-point or using a molded structure with thermal lamination, digital light processing is relatively fast and with the use of a digital mirror device (an array of up to several millions of mirrors which can be independently rotated to an on- and off-state), partial and entire layers of a scaffold can be simultaneously photopolymerized via projection [30–32]. The ability of such approach to fabricate scaffolds capable of supporting cell proliferation and differentiation was investigated by producing hydrogels from photocrosslinkable PEG diacrylates. By altering controlled release particles or biofactors within the polymerizable resin, each layer or partial layers were constructed from various controlled release microparticles thereby generating spatially distributed environments with microsize resolution. Moreover, digital mirror device can also be used to create complex and precise internal architectures such as pore shape and size. Murine bone marrow-derived cells were successfully encapsulated or seeded on patterned scaffolds functionalized with fibronectin.

Later, the study by Kim et al. also suggested due to the accurate and rapid printing speed, cells were not damaged during the printing process and in vitro observations revealed cells proliferated and distributed evenly over the hydrogel [33]. In addition, the in vivo transplantation of 3D-printed hydrogels using digital light processing approach into partially defected trachea rabbit model demonstrated the possibility of implementing digital light processing bioprinting in tissue fabrication [34].

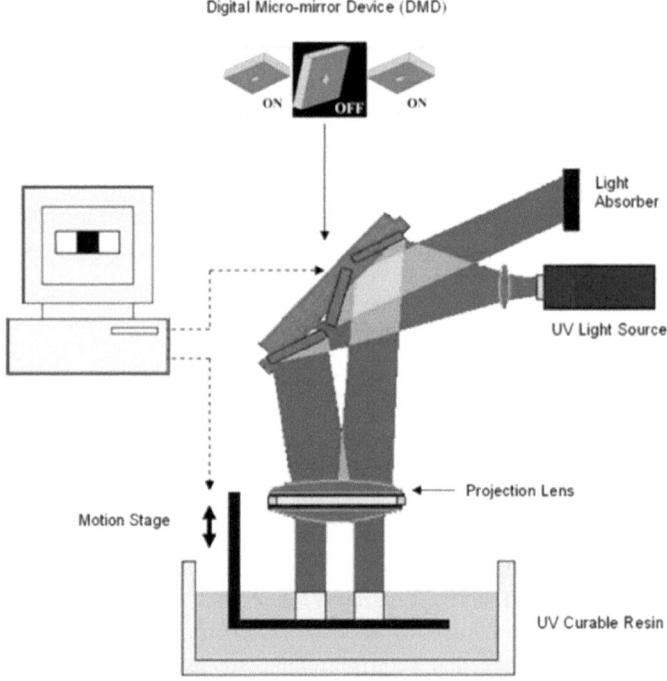

Fig. 2.4 Schematic of the digital micromirror device microstereolithography (DMD-μSL) setup. Reprint with permission from [30]

Despite the high production costs, the rapid printing speed of digital light processing-based 3D printing has created possible opportunities and enabled research and development into areas such as microfluidics [35], artificial skin model [36], drug delivery system [37], and tissue regeneration [38–42]. The study by Koffler et al. attempted to fabricate a complex central nervous system structure using a microscale continuous projection printing approach [38]. The authors claimed with such an approach, 3D biomimetic hydrogel scaffold based on PEG diacrylate and gelatin methacrylate matching the dimensions of the rodent spinal cord can be fabricated in under 2 s and can be scalable to human spinal cord sizes and lesion geometries.

2.1.4 Inkjet-Based Approach

This approach, which started as 2D-based inkjet printer for texts and images, has been used as a free-form fabrication method for the construction of 3D components as well as for producing arrays of nuclei acids and proteins [43]. Inkjet is a non-contact technology in which droplets of ink are jetted directly to a predefined location from a small aperture. Inkjet printing can be classified into either continuous inkjet printing or drop-on-demand printing [44]. During continuous inkjet printing, an electric charge is applied to the drops selectively as they formed out of the continuous ink stream once the drop break-off mechanism is activated. Uncharged drops would proceed directly onto the media to create an image while charged drops are deflected into a reservoir for recirculation when they passed through the electric field.

The development of a drop-on-demand inkjet printing was also gaining popularity and it is the preferred option for the fabrication of biological structures for soft tissue engineering applications as the process eliminates the inherent unreliability of the recirculation system used in continuous inkjet technology and the concerns regarding the possibility of contamination [1]. Furthermore, the use of complex drop charging and deflection hardware is not required for drop-on-demand printing [44]. Liquid droplets are ejected using either thermal or piezoelectric force. In thermal inkjet printers, liquid is vaporized via pulses of pressure generated by electrical heating. Tiny droplets are then ejected from the nozzle using pulses of air pressure. The temperature can reach between 200 and 300 °C and this could result in denaturalization of hydrogels or biocomponents in hydrogels. In piezoelectric inkjet printers, piezoelectric actuator is used to create pressure once an external voltage is applied and this pressure is then used to eject droplets from the nozzle [1]. It should be mentioned that both continuous inkjet and drop-on-demand systems are capable of printing with droplet that ranges in sizes between 15 and several hundred microns [1].

Based on the conventional inkjet printing process, inkjet-based bioprinting can deposit picolitre droplets of "bioink" precisely under computer control onto a hydrogel substrate [45]. For bioprinting, the piezoelectric approach in comparison with thermal inkjet printer enables droplets to remain directional with regular and equal size [45]. The study by Nakamura et al. demonstrated living cells were able to be ejected safely onto culture disks [46]. Later, a study was carried out to examine the behavior of cells exposed to the fluid and mechanical stresses associated with the use of drop-on-demand inkjet printing system utilizing piezoelectric approach [47]. Observations suggested that it is possible to alter the stresses experienced by the cells through the fluctuation of the amplitude and rise time of the electrical pulse used to excite the piezoelectric actuator. Furthermore, the amplitude of the pulse had a small influence on cell survivability (from 98 to 94%), while the rise time had no effect on the survival of cells. Post-printing, cell viability was also investigated and the cells that survived were untouched by the printing process, with neither pulse rise time nor amplitude causing any noticeable influence on cell viability. On the other hand, as stated in the study by Saunders et al., inkjet printing requires cell suspensions to be stable over several minutes during the printing process and they discovered after approximately 20 min, some cell agglomeration or sedimentation influenced the printing performance [47].

It has also been suggested that bioink with a viscosity of less than 10 mPa/s can only be produced using the inkjet printing approach [5]. The size of the deposited droplets can be regulated from 1 to 300 picolitres with the deposition rates varying from 1 to 10,000 droplets per second [48]. Nevertheless, the droplets must reach solidification into their predefined geometry before the next layer of droplets is deposited. Accordingly, it has been suggested that factors such as rapid gelation and the ability of the dispensed droplets to retain a 3D structure are crucial for the application of inkjet printing to fabricate high-resolution 3D hydrogel structures [49].

Reactive inkjet printing is based on the use of drop-on-demand with piezoelectric actuators to dispense ink droplets onto a substrate. However, reactive inkjet printing transforms the printing process from being simply a deposition approach to one where tiny droplet of one reactant can be added to another permitting reactions to be carried out at a small scale with precision [50, 51]. The combination of one droplet of reactant with the droplet of a second reactant using an inkjet printer is the principal notion behind reactive inkjet printing [50]. Reactive inkjet printing can be categorized as either single reactive inkjet printing where an inkjet printer deposits a reactant on top of another reactant deposited using a separate approach or full reactive inkjet printing in which the printer is able to dispense more than one type of ink, thus permitting the deposition of two or more reactants [50].

Studies have been carried out to examine the fabrication of hydrogels 3D printed using reactive inkjet printing [51, 52]. Duffy et al. attempted to print polymer hydrogels based on poly-ε-lysine, gellan gum, and natural peptides and printing alternating layers of ink created a distinct surface pattern based on the immediate formation of ionic bonds between polymers of opposing charges [51]. The study by Teo et al. attempted a technique that centered on the collision between the crosslinker and gelator prior to impacting the substrate surface and the intention was to create a scenario where hydrogel synthesis and patterning can occur in a single process [52]. They further commented that conventional inkjet printing of hydrogel is in essence restrained by the low viscosity of the printed ink, indicating crosslinking of hydrogel must occur after printing. This could be problematic for hydrogels as the slow diffusion of the crosslinking species through the gel results in very slow vertical printing speeds and this could result in dehydration of the gel and cell death if both are deposited simultaneously. The use of reactive inkjet printing allowed for a significant increase in vertical printing speeds. The droplet collision was demonstrated to increase advective mixing prior to impact, reducing the time need for gelation to take place, and improving definition of printed patterns.

2.2 4D Printing

The development of 4D printing began in 2014 when Tibbits described the manufacture of constructs using smart materials [53]. 4D printing can be defined as the fabrication of objects using the combination of additive manufacturing, stimuli-responsive materials, and to a certain degree mathematical modeling, which is essential to accurately predict shape-morphing behavior over time [54].

According to Momeni and Ni [54], these three components create an encoded static structure that becomes a dynamic smart structure through an interaction mechanism under an appropriate stimulus. Moreover, they commented the fourth dimension in 4D printing indicates predictable and desired evolution over time that depends on 3D space, and this 3D space is the specific arrangement of (physical) voxels determined by mathematics and realized by printing. External stimuli in 4D printing are essential activators for creating the planar output which is fabricated using a 3D printer to create a planned structure over time. The time-dependent characteristics of structures manufactured using 4D printing lie on the special spatial arrangement between passive and active materials in the 3D space arranged by the printing process [54].

In 2016, Ge et al. further hypothesized that 4D printing could create high-resolution, multimaterial shape memory polymer architectures [55, 56]. Typically, shape memory materials are needed for 4D printing and these materials must be able to be used in an optimized 3D printer to facilitate alterations in their physicochemical characteristics when exposed to selected stimulation [57]. Presently, smart materials available are limited since 4D printing is still at its infancy and majority of them are either hydrogels or shape memory polymers [58]. Briefly, shape memory polymers are materials that can hold temporary shapes and upon exposure to external stimulus, such as heat, will recover and return to its permanent form [55].

It has been demonstrated that hydrogels will undergo swelling as soon as solvent molecules diffuse or penetrate the polymer network and vice versa experience shrinkage once solvent molecules exit the network [55]. Consequently, hydrogels can alter its dimensions anisotropically and sizably just by regulating the magnitude and rate of shrinkage and/or swelling smartly. Hence, it is possible to print hydrogel architectures with programmable shape transformation capacity using 4D printing process [58]. However, hydrogel-based 4D printing has been suggested to be unsuitable for actuation and structural applications due to their low modulus and the slow rate of response from solvent diffusion [55].

Currently, several production techniques are being examined for their potential as 4D printers, and essentially, they are all based on the same additive manufacturing techniques used in 3D printing such as digital light processing, extrusion-based printing, and stereolithography [58, 59]. As described earlier, even though the same 3D printing technique is used for 4D printing, what sets 4D printing apart is the property of the materials used [59]. The ability to create structures that are dynamic in nature is the primary advantage of 4D printing in comparison with 3D printing. It has been well established that several factors will influence the process of 4D printing and one of the most vital factors is the stimulus-material relationship as it is the responsive material that introduces the fourth dimension into the process [59, 60].

Stimulus-responsive hydrogels can be further categorized into three groups based on the type of stimulus such as temperature or light (physical stimuli) or changes in pH (chemical stimuli). In addition, there are hydrogels that respond to the presence of elevated levels of glucose or respond upon detection of a target molecule such as an enzyme [61, 62]. They can be prepared via the design of the polymeric molecular chains. The physicochemical and structural properties of the hydrogels can be controlled by changing the external environmental stimulation such as pH and temperature; this in turn alters the degradation and swelling characteristics of the hydrogel [61, 62]. More importantly, it is vital to design and synthesize hydrogels to match the corresponding external stimulus environments. Stimulus-responsive hydrogels can quickly experience a volume phase transformation of discontinuous change once a critical point is reached due to changes in these external factors such as temperature or pH [62].

2.2.1   Thermo-Responsive

Based on the accepted averaged normal body or physiological temperature of 37 °C, thermoresponsive or thermosensitive hydrogels have gained considerable interest, particularly in the medical arena and several mechanisms have been developed to regulate the temperature difference from room temperature in vivo [61]. Once a thermoresponsive hydrogel reaches a certain temperature, the phase transition properties will trigger a response that could generate a change in affinity to the solvent. This is known as the critical solution temperature [62]. Tiny adjustments in temperature around a critical solution temperature render the chains extend or collapse, reacting to changes in the hydrophilic and hydrophobic interactions between the aqueous medium and the polymer chains [63]. Furthermore, thermoresponsive hydrogels can be divided into either positively or negatively responsive systems and these systems are typically identified by possessing an upper critical solution temperature (UCST) or a lower critical solution temperature (LCST), respectively [61].

One of the most extensively used and investigated temperature-responsive hydrogels are poly(N-isopropylacrylamide) (P(NIPAAm)) hydrogel and the LCST response of this reverse thermoresponsive hydrogel at approximately 32 °C in an aqueous medium is due to the equilibrium between the hydrophilic and hydrophobic groups as well as the efforts of enthalpic and entropic needed to solvate these groups [61, 64]. Below the lower critical solution temperature, P(NIPAAm) hydrogel displays a random coil structure (hydrophilic state) and a collapsed globular structure above the lower critical solution temperature (hydrophobic state). Moreover, this hydrogel absorbs water and exists in swollen states below the lower critical solution temperature, but will undergo an abrupt and drastic shrinkage in volume once the temperature of the medium is heated to a temperature above the lower critical solution temperature [64].

Despite its extensive investigations in drug delivery and in tissue engineering, P(NIPAAm) hydrogels are primarily produced using techniques such as lithography and molding that are fundamentally restricted to a 2D space [65]. A study was attempted in 2018 to 3D print P(NIPAAm) hydrogel using a high-resolution projection microstereolithography. The authors stated that the temperature-dependent deformation of the 3D-printed P(NIPAAm) hydrogel could be controlled by adjusting the polymer resin composition and the manufacturing process parameters [65].

Due to its capacity to process various ink materials as well as its ease of operation, micro-extrusion-based 3D printing is one of the most widely used 4D printing technologies [58]. A hydrogel consisted of P(NIPAAm) precursors, sodium alginate, and methylcellulose was developed and its potential use as wound dressing materials was examined. PEG diacrylate was added as a hydrophilic co-monomer in addition to serving as the long-chain crosslinker of P(NIPAAm). The micro-extruded hydrogel was then loaded with a mixture of octenidine dihydrochloride and 2-phenoxyethanol, rendering the hydrogel antimicrobial. The authors further commented that the temperature-stimulated polymer was able to sense the temperature locally and regulated the swelling ratio and in vitro drug diffusion as well as demonstrating shape-morphing behavior [66]. It has also been suggested that the integration of microfluidic systems with conventional extrusion-based 3D printing would allow the precise control of the structural and compositional properties of tissue engineering scaffolds [67]. Recently, a study was attempted to develop MXene-incorporated hydrogel scaffold using a microfluidic-assisted 3D printing approach for skin flap regeneration. Utilizing the photothermal conversion capacity of MXene nanosheets and temperature-responsive ability of P(NIPAAm) hydrogels, the scaffold was said to display a near-infrared responsive swelling/shrinkage behavior. The study further postulates the scaffold could promote proliferation, migration, and proangiogenic effects of endothelial cells under near-infrared irradiation by incorporating vascular endothelial growth factor into the hydrogel matrix. In vivo animal study suggested the scaffold could effectively improve skin flap survival by promoting angiogenesis, reducing inflammation, and attenuating apoptosis in skin flaps [67].

Although being less common than LCST systems, interpenetrating network and other hydrogels that display UCST behavior normally centered on an acrylic acid-based monomer, and similar to LCST system, the acrylic acid networks can be altered with hydrophilic or hydrophobic co-monomers to shift the critical temperature [61]. A study was attempted to 4D print hydrogels based on poly(acrylic acid) that could possess self-healing and shape memory properties simultaneously. Using a stereolithography-based resin printer, hydrogel printing was achieved through solvent-free copolymerization of the hydrophilic acrylic acid and hydrophobic hexadecyl acrylate monomers in the presence of photoinitiator followed by swelling in water [68]. In addition, the melting and crystallization temperature enabled the printed hydrogel to possess a shape memory effect near the human physiological temperature. The authors hypothesized the self-healing and shape memory properties could be triggered near the body temperature by altering the molar ratio of the monomers [68].

2.2.2   pH-Responsive

pH-responsive hydrogels consist of ionic pendent groups within polymeric backbones. The pendant groups will ionize and generate fixed charges on the polymer network once they are exposed to an aqueous medium with appropriate pH and ionic strength. This results in the creation of electrostatic repulsive forces and they are responsible for the pH-dependent swelling and shrinkage of the hydrogel [61, 69]. Minute changes in pH could result in notable change in the mesh size of the polymeric networks. Two different classes of pH-responsive hydrogel are available and they differ in their pendant group ionization and the resultant swelling characteristics. In anionic hydrogels, the pendant groups are ionized above the acid dissolution constant (pK_a) and are un-ionized below the pK_a value. If the hydrogel is exposed to a medium that has a pH value greater than its pK_a value, it will swell due to the large osmotic pressure driven by the availability of ions. On the other hand, the pendant groups in cationic hydrogels are ionized at a pH less than its pK_a value, and subsequently, the hydrogel will swell in mediums that have a lower pH than the pK_a value [61, 69].

Furthermore, several factors will regulate the magnitude of swelling of ionic hydrogels and these include the properties of the swelling medium and the properties of the hydrogel such as pK_a, concentration, and charge of the ionizable group. The pH and ionic strength of the solution will determine the composition of the swelling medium, which is regulated by the primary counterions and their valency in solution [61, 69].

Bacterial infections associated with the use of implantable devices and prosthetics are complications that occur frequently and by far the most common. The primary issue related to the application of antibiotics is to ensure its activity, and release is maintained for an extended period post-operation. Furthermore, high dosage of certain antibiotics has been demonstrated to be nephrotoxic and ototoxic as reported by previous studies. The loaded dosages for most controlled release systems are generally high and for that reason, the systemic exposure of antibiotics in urine and blood is a primary safety concern. Hence, a more appropriate form of treatment is to deliver antimicrobial agents locally and in a targeted manner [70]. Garcia et al. investigated the fabrication of pH-responsive and antimicrobial hydrogels 3D printed using stereolithography. Acrylic acid is one of the most common monomers used to introduce the pH-responsive behavior. According to the authors, the printed hydrogel demonstrated reversible shrinkage and swelling upon environmental changes on the pH and the extent of swelling/shrinkage was directly related to the amount of acrylic acid. Using *Staphylococcus aureus* as a bacterial model, the antimicrobial properties of the hydrogels were examined [71].

pH-responsive hydrogels, due to their high-volume response to environmental pH changes, have been extensively investigated for the controlled delivery of drugs and therapeutics. The controlled drug release from these loaded hydrogels is triggered by a change in the surrounding pH such as during the transit through the gastrointestinal tract [61]. A recent study by Wang et al. proposed the fabrication of a hydrogel drug carrier using pH-responsive materials and semi-solid extrusion 3D printing, thus enabling site-targeted drug release and the possibility to customize the temporal release profiles. The swelling properties of the hydrogel were examined under both artificial gastric and intestinal fluids, and the authors claimed high swelling rates at either acidic or alkaline conditions could be achieved through the adjustments in the mass ratio between sodium alginate and carboxymethyl chitosan. They also claimed controlled release could be achieved by tuning the infill density of the printing process [72].

2.2.3 *Magneto-Responsive*

The possibilities of utilizing a magnetic field to trigger as well as controlling the properties of magneto-responsive hydrogels have also been examined. These hydrogels are typically consisted of a base hydrogel and a magnetic component such as magnetic iron oxides, maghemite, and magnetite, which renders the resultant hydrogel sensitive to external magnetic field [73–75]. As an external stimulus, a magnetic field is perfect due to its precise and quick field-induced controllability [74]. It has been suggested that the ability to activate remote actuation with a quick response rate and biocompatibility even at high field strength makes electromagnetic an ideal stimulus particularly for in vivo applications [76].

Magnetic particles will gather immediately once an external magnetic field is applied, and this will lead to solvents being forced out of the hydrogel network as it contracts. Ultimately, the shape of the hydrogel will change rapidly [62]. The concentration, size, composition, and how uniform the magnetic particles are in the hydrogel will determine the properties of the magneto-responsive hydrogel [62]. One of the frequently used magnetic materials for biomedical application such as tissue engineering is superparamagnetic iron oxide nanoparticles [73–75]. However, clustering of magnetic nanoparticles can dramatically alter their collective magnetic properties, and superparamagnetic iron oxide nanoparticles tend to aggregate because of interparticle magnetic interactions resulting in the so-called multicore particles [75, 77]. The possibility of these nanoparticles and multicore particles to aggregate and sediment in neutral pH is one major disadvantage [75].

In 2019, a study hypothesized dynamic tissue scaffold that can imitate the extra-cellular matrix environment within the human body could be fabricated using 3D printing by combining hydrogels fabricated from oxidized sodium hyaluronate and glycol chitosan with superparamagnetic iron oxide nanoparticles. The study suggested dimensional changes in the scaffold could be achieved through the application of a magnetic field. Additionally, in vitro observation showed the chondrogenic differentiation of ATDC5 cells encapsulated and cultured within the hydrogel was significantly affected by the strength of the magnetic field [73].

2.2.4 Moisture/Humidity-Responsive

First applied as a stimulus in 4D printing, hydrogels sensitive to humidity or moisture can deform and revert to its original form once dried [78]. Water-responsive hydrogels or sometimes referred to as superabsorbent polymers are predominately composed of acrylic monomers and are weakly crosslinked. For that reason, they display an extremely high liquid swelling capacity [76]. Moreover, renewable raw materials such as cellulose have also been examined [79].

It is well known that the degree of crosslinking will influence the swelling ratio of hydrogel, as highly crosslinked structures will have a lower swelling ratio than a hydrogel with a relatively lower crosslinking density [80]. Lv et al. examined the potential of humidity-driven movement of PEG diacrylate hydrogel microstructures fabricated using femtosecond laser direct writing. By adjusting the fabrication parameters, the voxels of hydrogel microstructures possess controllable crosslinking density and hence controllable humidity-driven swelling ability. They claimed the humidity-responsive actuation of hydrogel microstructures was repeatable and stable over ten thousand cycles [81].

Recently, a study was attempted to create a device that could utilize the swelling properties of a water-absorbent hydrogel to trigger a self-folding mechanism upon exposure to bodily fluids [82]. According to the authors, this folding mechanism was achieved by printing a bilayer of a flexible polyurethane printing resin and a highly swelling sodium acrylate hydrogel using photopolymerization. Their intention was to develop a cuff electrode for small-nerve interfacing and the hydrogel functions as the actuation mechanism that swells once it is in contact with fluid after the device is implanted (Fig. 2.5) [82]. Similarly, the study by Yang et al. also explored the swelling properties of hydrogels and the shape-changing mechanism once exposed to an aqueous medium. Their intent was to utilize 4D printing of a gelatin hydrogel to fabricate a bundle consisting of cell-laden fibers mimicking the natural structure of the skeletal muscle tissue [83]. According to the authors, one of the most vital aspects inducing folding was the grooves on the gelatin films. The films initiated swelling when exposed to water and the thickness of the film changed after swelling while the length and width remained constant. Due to the grooved topology, different swelling phenomena occurred on the surface and bottom of the film. This caused the gelatin film to "roll up" forming a tubular structure (Fig. 2.5).

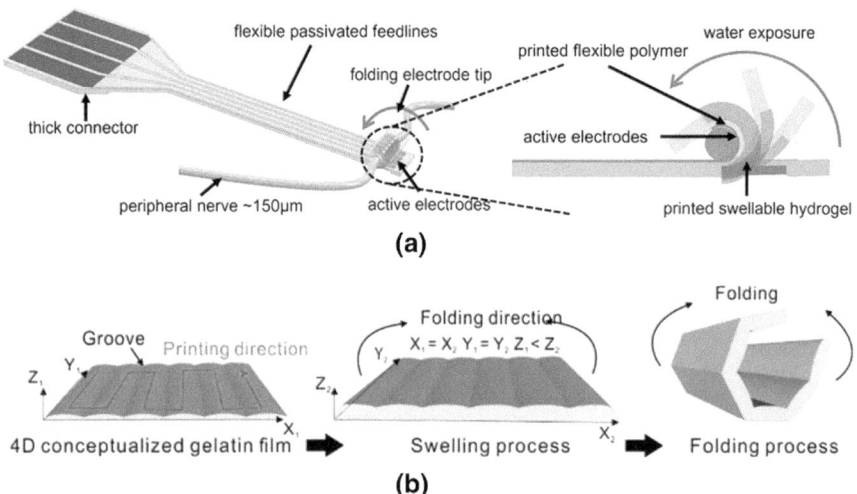

Fig. 2.5 Shape morphing mechanisms of moisture-responsive hydrogel. **a** 4D-printed self-folding electrode. Modified from [82]. http://creativecommons.org/licenses/by/4.0/. **b** Shape morphing mechanism of the gelatin film and one of the most important factors inducing folding are the grooves on the gelatin films. Modified from [83]. http://creativecommons.org/licenses/by/4.0/

It should be mentioned that according to Li et al., the constantly high humid environment during in vitro cell culture experiments and in vivo implantations would restrict the transformation capability of humidity or moisture-responsive hydrogels to once only [84]. More importantly, specific osmotic pressure is needed for cells to function normally [85]. Consequently, constraints in the humidity or osmotic pressure that can be applied may restrict the magnitude of shape changing in humidity-responsive hydrogels.

2.3   3D Bioprinting

During 3D bioprinting, both cells and biomaterials are deposited simultaneously with micrometer precision to create tissue-like structure and the process has shown potential in tissue engineering and regenerative medicine for the fabrication of a variety of transplantable tissues such as bone, cartilage, and skin. Numerous advantages are associated with the use of 3D bioprinting such as high-resolution cell deposition and accurate control of cell distribution. Bioprinting, distinct from conventional 3D printing approaches discussed earlier, requires a different technical methodology that is compatible with depositing living cells (Fig. 2.6) [86].

Deposition approaches such as laser-induced forward transfer (LIFT) and extrusion and inkjet-based printing have been studied for the biofabrication of hydrogels as a carrier for cells and/or bioactive compounds, and each deposition approach

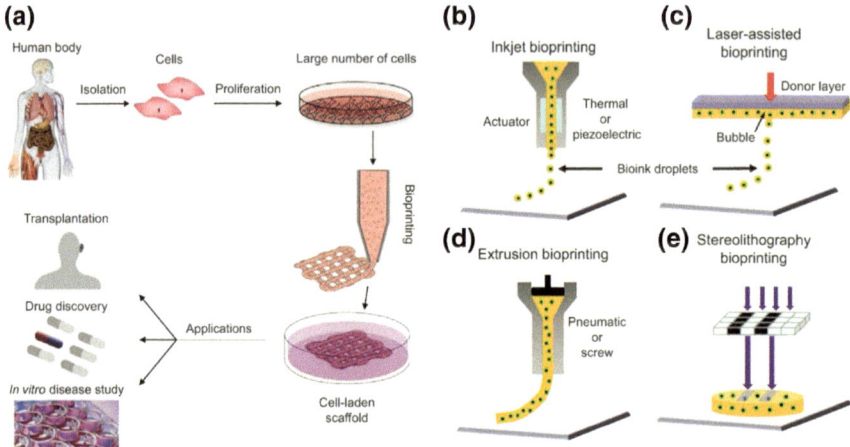

Fig. 2.6 3D bioprinting for human therapeutic applications. **a** Typical workflow including cell isolation and expansion prior to printing the desired cell-laden scaffold. **b–e** Various bioprinting techniques are available. Reprint with permission from [86]

requires very specific prerequisites to cater for the properties of the hydrogel-based bioinks used such as their post-curing rate and rheology. This is to ensure the reliable fabrication of 3D structures [87]. However, despite their use in bioprinting, no single technique allows the biofabrication of all complexities and scales of synthetic tissues, and each methodologies have its own disadvantages [86–88]. For example, only a handful of laser-assisted bioprinters are available due to their high cost as well as their complexity in comparison with other types of printers [86]. Likewise, the use of inkjet printers could cause significant cell damage and death in addition to cell sedimentation and aggregation as a consequence of small orifice diameter which restricts printing cells in high densities [87, 88]. Similarly, extrusion-based approaches can induce cell deformation due to shear stress or generate larger pressure drops at the nozzle that can potentially be harmful for embedded cells [87, 88].

In terms of choosing the correct cells for bioprinting, the key issues that need to be considered are to what extent the bioprinted cells can develop or retain their functions under optimized microenvironments in vivo in addition to how closely these printed cells can imitate the physiological state of cells in vivo. In an ideal situation, patient's own cells should be used for bioprinting to avoid any negative immune responses. Stem cells are the most promising cell source as not all types of cells can regenerate after damage (Fig. 2.6) [86].

2.3.1 *Bioinks*

According to the study by Lee et al., suitable bioinks are needed for biofabrication and cell printing, as they must permit appropriate diffusive transport of oxygen, nutrients, and metabolic wastes between the culture medium and embedded cells while at the same time shielding the embedded cells from the external mechanical stresses during the printing process [89]. In addition, the bioink must be non-cytotoxic and biocompatible and allow the laden cells to attach and proliferate on top of providing a 3D structure that is mechanically stable and can maintain structural integrity.

The total number of bioinks needed for a given print can be reduced if stem cells are used, but will increase the level of complexity to the biofabrication process. It has also been suggested that the availability of addition bioink formulations containing various biological materials such as growth factors to assist in guiding site-specific differentiation would be advantageous. Then again, post-printing culture will be an issue even without utilizing this approach as growth factors and differentiation stimulators need to be deposited in a precise manner in an effort to ensure the differentiation process is regulated particularly when vascularization is desired [86].

The characterization and production of new bioinks have attracted an increasing amount of attention in recent years due to the lack of materials suitable for bioprinting, and this inadequacy was considered as a major issue limiting the advancement of biofabrication and cell printing technology [89, 90]. First proposed by Mironov [91] and later in the review by Groll et al. [90], bioink is a term first used in 2003 in the context of organ printing and the concept initially was to print a hydrogel and tissue spheroids or living cells were then inserted as the "bioink" through bioprinting. Consequently, the term bioink originally referred to the cellular component that was situated in 3D on or within hydrogels. More importantly, biomaterials that can be considered as a bioink must function as a medium for cell delivery during formulation and processing. For further distinction and clarification, Groll et al. proposed that (bio)materials that can be printed and later seeded with cells after printing but not formulated directly with cells do not qualify as a bioink but instead these materials should be classified as biomaterial inks [90]. These biomaterial inks could be utilized in conjunction with bioink-fabrication in hybrid approaches to fabricate an intrinsic mechanical support within the construct [90]. They further proposed that bioinks should be defined as "a formulation of cells appropriate for processing through an automated biofabrication technology that may also contain biologically active components and biomaterials." Moreover, bioinks may include cells in various environments and forms such as single cells, encapsulated in tailored colloidal microenvironments, cellular rods, cells seeded onto microcarriers, cells organized in minitissues organoids, cells aggregated in spheroids, or cells coated by a thin layer of material. Although not a necessity, but bioinks can also contain bioactive molecules such as DNA, miRNA, cytokines, exosomes, growth factors, or biomaterials. Of vital significance is that this proposed definition of bioinks is independent to the technology used for biofabrication [90].

Hydrogels, as a well-known component of bioinks, are frequently used in biofabrication and cell printing technology based on the fact they can provide ideal extracellular microenvironment for different cell functions of encapsulated cells in 3D structures such as proliferation and differentiation [4]. Natural hydrogels such as collagen, gelatin, alginate, chitosan, fibrin, silk, and hyaluronic acid are extensively utilized in bioprinting research, as cells will benefit from the availability of abundance of chemical signals in these hydrogels leading to high viability and proliferation rates. Yet, construct reproducibility is often difficult based on batch-to-batch variation and the sensitivity of stem cells to these variations. On the other hand, synthetic hydrogels such as PEG can be used to fabricate 3D-printed structures with high shape conformity but often with low cell viability as these structures provide an inert environment for embedded cells without active binding sites. Subsequently, bioactive compounds such as growth factors need to be grafted or added to the hydrogel to enhance regulation over cellular differentiation [87].

In addition, there are certain challenges posed by the use of hydrogels for 3D tissue printing as stated in the study by Cohen et al. [92]. Regulating the sol–gel transition is necessary using 3D printing so that the hydrogel is viscous enough to maintain its shape when patterned on a substrate, while at the same time it needs to be sufficiently inviscid to preserve mechanical integrity when being extruded out of a print nozzle despite the shear forces. Above all, cell viability must also be maintained throughout the printing process. This can be achieved by eliminating the use of cytotoxic initiators during the sol–gel transition as well as reducing the amount of shear forces acting on cells during deposition.

Prior to printing, the crosslinking abilities, gelation, and viscosity of the hydrogel-based bioink are some of the properties that need to be considered as they can drastically influence the viability, proliferation, and morphology of cells after printing. It is a challenge to fabricate a hydrogel that can simultaneously provide a structurally secure scaffold and protect and support cells. This is because different mechanical requirements are associated with these functions [4, 87, 93].

2.3.2   Cell Viability and Structural Integrity

In 2013, Malda et al. introduced the concept of biofabrication window [87]. The main issue is that the printing of complex, high-resolution tissue-like constructs requires narrow boundaries for the physical properties of the hydrogel. On top of that, the construct should enable migration, proliferation, and differentiation of the embedded and endogenous cells. In the end, the process of 3D bioprinting inflicts contrasting requirements on the hydrogel and a balance must be reached between cell viability and structural integrity.

Increasing the crosslinking density has always been the traditional approach to enhance hydrogel printability. However, aqueous environments are best for cells to thrive where their migration is not restricted by the dense and stiff network. On the other hand, low crosslinking density will lead to a hydrogel with an inability to retain shape during printing. Subsequently, a compromise must be reached to achieve a hydrogel with sufficient mechanical properties and able to facilitate maximal cell and tissue compatibility [4, 87].

The physicochemical characteristics of the hydrogel such as crosslinking mechanism and rheological properties will determine its suitability for a certain biofabrication process as well as the conditions imposed by the instrument/printer. In addition, there are explicit processing factors that will govern the magnitude of shear stress the embedded cells will experience or how long would it take to produce a clinically relevant structure. Lastly, the printed structure must develop and maintain shape fidelity and adequate mechanical stability once the hydrogel precursors have been printed and the cells have survived [87].

2.3.3   Translational Applications

2.3.3.1   Tissue Constructs

It is well known that bone is a metabolically active tissue consisting of calcified regions and supplied by an intraosseous vasculature and without this vascular supply, nutrients are primarily transported by diffusion which is effective only over short distances or for tissues with low metabolic activity. More importantly, the lack of functional microvasculature connected to the host blood supply is a major concern given that once an engineered tissue construct is implanted, there will be a limited capacity for the seeded cells to both clear metabolism by-products and to uptake substrate molecules such as oxygen and glucose. These limitations impair cell viability and to a certain degree hinder the success of engineered tissue [94]. For that reason, engineering functional bone constructs requires the presence of distinct niches that can support vasculogenesis in addition to osteogenesis [95]. Byambaa et al. utilized an extrusion-based direct-writing approach to fabricate microstructured bone-like tissue constructs containing a perusable vascular lumen [95]. In their study, the authors intended to develop a bioactive gelatin methacryloyl bioink functionalized with vascular endothelial growth factor to illustrate the concept of producing vascularized bone tissue. A perfusable blood vessel within the bioprinted construct was created through the printing of a central cylinder composed of a rapidly degradable gelatin methacryloyl hydrogel. To induce osteogenesis, they further bioprinted around the central vascular fiber a gelatin methacryloyl hydrogel loaded with silicate nanoplatelets. The bioprinted constructs were then used as biomimetic in vitro matrices to co-culture human umbilical vein endothelial as well as bone marrow-derived human mesenchymal stem cells and their results revealed the construct was able to support cell survival and proliferation during maturation in vitro and displayed high structural stability for 21 days during the in vitro culture period.

Furthermore, stem cell-mediated bone repair has been used in clinical trials for the regeneration of large craniomaxillofacial defects and the successful regenerative outcomes in these investigations have provided a solid foundation for wider utilization of stem cells in skeletal repair therapy [96]. The integration of stem cells and 3D bioprinting shows promise in aiding in the reconstruction of cranial defects. Using a digital light processing approach, a study postulated that the in vivo implantation of a porous hydrogel could considerably improve the cell spreading and providing a cell-inspired microenvironment for preserving and enhancing the functions of encapsulated stem cells [97]. Porous hydrogels consist of bone marrow stem cells mixed with gelatin methacrylate/dextran emulsion were used to investigate its efficacy in treating full-thickness craniotomy defects created in the parietal bone of rats. According to the authors, the results showed the skull was reconstructed successfully using the cell-laden porous hydrogel, and histological evaluation showed large quantity of newborn bone lacunas was observed and the defect area was completely occupied and tightly integrated with the host bone tissues.

An extrusion-based approach was applied by Das et al. to investigate the possibility of biofabricating 3D tissue constructs using silk fibroin-gelatin bioink encapsulated with human nasal inferior turbinate tissue-derived mesenchymal progenitor cells. The printed constructs were allowed to crosslink via enzymatic crosslinking using mushroom tyrosinase and physical crosslinking via sonication. The influence of rheology, secondary conformations of the bioink, temporally controllable gelation, and printing parameters were examined to achieve maximum cell viability and multilineage differentiation of the encapsulated progenitor cells. The authors claimed that the progenitor cells were viable in the printed construct over a month in in vitro culture conditions as well as maintaining a stable 3D structure [98].

The fabrication of intricate structures with dimensions relevant for human tissues is possible through 3D bioprinting but our inability to identify appropriate bioinks capable of printing translationally relevant tissue with complex geometries remains unresolved. In addition, the utilization of bioprinting can also be restricted by the availability of materials that both expedite bioprinting procedures and support cell viability and function by providing tissue-specific cues [99–101]. A study by Skardal et al. described a hydrogel system based on hyaluronic acid and gelatin that could potentially offer tissue-specific biochemical signals and mimic the mechanical properties of in vivo tissues [99]. In their study, biochemical factors were provided by incorporating tissue-derived extracellular matrix materials, which include potent growth factors. To examine their theory, hydrogel bioinks were fabricated and used to bioprint primary liver spheroids in a liver-specific bioink to create in vitro liver constructs with high cell viability and measurable functional albumin and urea output. Later, a tissue-specific bioink consists of pancreatic extracellular matrix and hyaluronic acid methacrylate was used to fabricate a hydrogel in which the authors claimed that it was able to maintain islet cell adhesion and morphology in vitro through the Rac1/ROCK/MLCK signaling pathway and mimic the microenvironment of the pancreas [100]. Moreover, in vivo experiments suggested that the 3D-printed islet-encapsulated hydrogel increased insulin levels in diabetic mice and maintained blood glucose levels within a normal range for 90 days and secreted insulin rapidly in response to blood glucose stimulation. Additionally, the authors claimed the printed hydrogel was able to facilitate the attachment and growth of new blood vessels and increase the density of new vessels.

The feasibility of printing functional tissue for transplantation using a tissue-specific hybrid bioink was also examined in a study by producing a bioink that has been reinforced with extracellular matrix derived from decellularized tissue (rECM) [101]. The effect of angiogenesis in vivo was examined by subcutaneously implanting bioprinted hydrogel into T-cell deficient mice to mimic clinical immunosuppression in transplant patients. Based on their observations, the authors stated that the hydrogel appeared well integrated into the surrounding tissue without any obvious signs of foreign body response or inflammation. Furthermore, they also examined if a hydrogel derived from 3D-printed rECM but also containing extracellular matrix derived from lung tissue could support the growth and differentiation of primary epithelial progenitor cells isolated from normal human airways. Airway lumens remained obvious with viable cells for one month in vitro with evidence of differentiation into mature epithelial cell types found in native human airways. Another study attempted to print an artificial cornea with a high degree of precision, smoothness, and programmable curvature using digital light processing 3D bioprinting in one piece with no support structure [102]. Bioink derived from corneal decellularized extracellular matrix and gelatin methacryloyl was synthesized. In vitro experiments suggested the composite hydrogel could maintain high cell viability and expressed core proteins when loaded with human corneal fibroblasts. The in vivo safety and therapeutic potential of the composite hydrogel were examined in a focal corneal defect animal model with the corneal defects of 5 mm diameter and one-third the depth of the cornea and results indicated the hydrogel might promote epithelial regeneration and restore clarity.

Seeding cells on polymeric substrates or condensing cells in a plastic mold are two approaches developed for the creation of bioengineered cardiac tissues but their utilizations have been restricted based on a lack of cellular organization, uniformity, and scalability [103]. A study by Wang et al. attempted to overcome these limitations by applying a bioprinting approach to fabricate functional cardiac tissues with spontaneous synchronous contraction and spatial organization similar to native myocardium [103]. Primary cardiomyocytes isolated from infant rat hearts were suspended in a fibrin-based composite hydrogel bioink and sequentially printed to form uniformly organized, centimeter-scaled cardiac tissues. The authors claimed the bioprinted cardiac tissue has a spontaneous synchronous contraction in culture suggesting in vitro cardiac tissue development and maturation. This was based on their observation that the cardiac tissue was unlike native tissues both functionally and structurally immediately after bioprinting but instead they gradually matured after three to four weeks in culture. Immunostaining for α-actinin and connexin 43 indicated progressive cardiac tissue development and the tissues were formed with uniformly aligned, dense, and electromechanically coupled cardiac cells. In addition, the authors also claimed these printed tissues displayed physiologic responses to known cardiac drugs in relations to contraction forces and beating frequency.

2.3.3.2 Skin Repair

Instead of the traditional low precision cell spraying and seeding techniques, bioprinting has shown promise in the precise delivery of cells to replicate natural skin anisotropy [104]. Based on this notion, a bioink was composed of gelatin methacrylamide and collagen doped with tyrosinase was developed for 3D bioprinting of living skin tissues. In addition to being an enzyme to facilitate the crosslinking of gelatin methacrylamide and collagen, tyrosinase also functions as an essential bioactive compound in the skin regeneration process. The authors stated that the bioink was able to produce stable 3D living constructs. Cell culture experiments demonstrated that the human melanocytes, human dermal fibroblasts, and human keratinocytes displayed high cell viabilities. Based on the observations from in vivo animal tests, the authors further claimed that doped bioink could assist in the formation of an epidermis and dermis [105].

Mushroom tyrosinase was also used in another study as a crosslinking approach for a bioink that was examined for its potential to develop a human-cell-based full-thickness skin model [106]. Bioprinting was carried out using an extrusion-based approach and the bioink was composed of silk fibrin and gelatin with undulated morphology of epidermal rete ridges, architectural, mechanical, and biochemical functionalities similar to human skin. Human primary adult dermal fibroblasts were mixed with the bioink for the printing of the dermal layer. The epidermal structure was printed on top of the dermal construct using the bioink mixed with human adult keratinocytes after fibroblast proliferation and extracellular matrix deposition analysis were carried out [106].

A gelatin-alginate hydrogel was applied in a recent study to examine the feasibility of constructing a multilayer composite scaffold simulating the hair follicle microenvironment in vivo. Fibroblasts, human umbilical vein endothelial cells, dermal papilla cells, and epidermal cells were encapsulated in the composite hydrogel and respectively bioprinted into the different layers of the composite scaffold. The bioprinted scaffold with epidermis- and dermis-like structure was subsequently transplanted into full-thickness wounds in nude mice. The authors commented that the multilayer scaffold displayed suitable cytocompatibility and enhanced the proliferation ability of dermal papilla cells. It also facilitated the formation of self-aggregating dermal papilla cell spheroids and restored dermal papilla cell genes associated with hair induction. They further commented the dermal and epidermal cells self-assembled successfully into immature hair follicles in vitro. Based on the observations that hair follicles were regenerated in the appropriate orientation in vivo, the authors postulated that it was probably due to the hierarchical grid structure of the scaffold and the dot bioprinting of dermal papilla cells [107].

2.3.3.3 Cancer Therapy

Hydrogels offer an ideal solution in cancer immunotherapy and treatment due to the number of different materials systems that can be used. To determine the most suitable hydrogel-based system for a specific application, factors such as release kinetics, immunogenicity, and the site where the hydrogel will be used can serve as the selection criteria [108, 109]. Hydrogels have been used to deliver multiple types of immunotherapies based on proteins and cells. Recently, a study was attempted to examine the antitumor effect of natural killer cells encapsulated in micro/macropore-forming hydrogels using 3D bioprinting for tumor immunotherapy. Bioinks with appropriate rheological and shear-thinning properties for printing were fabricated by adjusting the ratio of alginate to gelatin. According to the authors, micropores were formed via the removal of thermally sensitive gelatin and macropores were formed through bioprinting process. The characteristics of pore-forming hydrogel, expansion with viability of immune cells in 3D culture, and functionality with an in vitro model were examined to confirm the possibility of immunotherapy. Observations suggested the hydrogel could enhance the activities and conditions of natural killer cells (NK92) such as viability, cytotoxicity, and cytokine release. The authors further postulate that the release of zEGFR-CAR natural killer cells from the hydrogel could result in lysis of the target cell when implanted at the tumor site, especially for solid tumors [110].

2.4 Concluding Remarks

As a result of its capacity to create complex, intricate, and highly customizable tissue engineering scaffolds that can promote cell infiltration and support adhesion, the fabrication of hydrogel constructs using 3D printing has become an attractive option for biomedical researchers. The notion behind the development of 4D printing began in 2014 and the process can be described as a fabrication approach that combines the use of additive manufacturing, stimuli-responsive materials and to a certain degree mathematical modeling. Currently, there are several production techniques being developed and examined for 4D printing, and in essence, they are all based on the same additive manufacturing techniques used in 3D printing. The ability to create structures that are dynamic in nature is the primary advantage of 4D printing in comparison with 3D printing. It has been well established that several factors will influence the process of 4D printing and one of the most vital factors is the stimulus-material relationship. This is because the responsive material is responsible for the introduction of the fourth dimension into the printing process. Bioprinting, dissimilar to conventional 3D printing, will need a different technical approach that is compatible with the deposition of living cells as the intention of bioprinting is to fabricate tissue-like constructs using small entities of biomaterials and cells dispensed with micrometer precision. Numerous advantages are associated with the use of 3D bioprinting such as high-resolution cell deposition and accurate control of cell distribution. However, no single bioprinting technique enables the biofabrication of all complexities and scales of synthetic tissues and each methodologies have its own disadvantages. Choosing the correct cells for printing as well as suitable bioinks are necessary for biofabrication and cell printing, as they must permit appropriate diffusive transport of oxygen, nutrients, and metabolic wastes between the culture medium and embedded cells while at the same time shielding the embedded cells from the external mechanical stresses during the printing process. The use of bioprinting has shown great potential in regenerative medicine and tissue engineering and observations from both in vitro and limited in vivo studies demonstrated the possibility of bioprinting and transplanting artificial organs in the future.

References

1. Jang TS, Jung HD, Pan HM et al (2018) 3D printing of hydrogel composite systems: recent advances in technology for tissue engineering. Int J Bioprint 4:126. https://doi.org/10.18063/IJB.v4i1.126
2. Chia HN, Wu BM (2015) Recent advances in 3D printing of biomaterials. J Biol Eng 9:4. https://doi.org/10.1186/s13036-015-0001-4
3. Wu GH, Hsu SH (2015) Review: polymeric-based 3D printing for tissue engineering. J Med Biol Eng 35:285–292
4. Li J, Wu C, Chu PK et al (2020) 3D printing of hydrogels: rational design strategies and emerging biomedical applications. Mater Sci Eng R Rep 140:100543. https://doi.org/10.1016/j.mser.2020.100543
5. Kaliaraj GS, Shanmugam DK, Dasan A et al (2023) Hydrogels—a promising materials for 3D printing technology. Gels 9:260. https://doi.org/10.3390/gels9030260
6. Billiet T, Vandenhaute M, Schelfhout J et al (2012) A review of trends and limitations in hydrogel-rapid prototyping for tissue engineering. Biomaterials 33:6020–6041
7. Song J, Michas C, Chen CS et al (2020) From simple to architecturally complex hydrogel scaffolds for cell and tissue engineering applications: opportunities presented by two-photon polymerization. Adv Healthc Mater 9:e1901217. https://doi.org/10.1002/adhm.201901217
8. Weiss T, Hildebrand G, Schade R et al (2009) Two-photon polymerization for microfabrication of three-dimensional scaffolds for tissue engineering application. Eng Life Sci 9:384–390
9. Liska R, Schuster M, Inführ R et al (2007) Photopolymers for rapid prototyping. J Coat Technol Res 4:505–510
10. Kufelt O, El-Tamer A, Sehring C et al (2015) Water-soluble photopolymerizable chitosan hydrogels for biofabrication via two-photon polymerization. Acta Biomater 18:186–195
11. Yuan Y, Chen L, Shi Z et al (2022) Micro/nanoarchitectonics of 3D printed scaffolds with excellent biocompatibility prepared using femtosecond laser two-photon polymerization for tissue engineering applications. Nanomaterials 12:391. https://doi.org/10.3390/nano12030391
12. Ostendorf A, Chichkov BN (2006) Two-photon polymerization: a new approach to micromachining. Photonics Spectra 40:72
13. Serra P, Fernández-Pradas J, Berthet F et al (2004) Laser direct writing of biomolecule microarrays. Appl Phys A 79:949–952
14. Serra P, Piqué A (2019) Laser-induced forward transfer: fundamentals and applications. Adv Mater Technol 4:1800099. https://doi.org/10.1002/admt.201800099
15. Fernández-Pradas JM, Serra P (2020) Laser-induced forward transfer: a method for printing functional inks. Crystals 10:651
16. Chang J, Sun X (2023) Laser-induced forward transfer based laser bioprinting in biomedical applications. Front Bioeng Biotechnol 11:1255782. https://doi.org/10.3389/fbioe.2023.1255782
17. Yusupov V, Churbanov S, Churbanova E et al (2020) Laser-induced forward transfer hydrogel printing: a defined route for highly controlled process. Int J Bioprint 6:271. https://doi.org/10.18063/ijb.v6i3.271

18. Gruene M, Deiwick A, Koch L et al (2011) Laser printing of stem cells for biofabrication of scaffold-free autologous grafts. Tissue Eng Part C Methods 17:79–87
19. Kryou C, Theodorakos I, Karakaidos P et al (2021) Parametric study of jet/droplet formation process during lift printing of living cell-laden bioink. Micromachines 12:1408. https://doi.org/10.3390/mi12111408
20. Barron JA, Rosen R, Jones-Meehan J et al (2004) Biological laser printing of genetically modified *Escherichia coli* for biosensor applications. Biosens Bioelectron 20:246–252
21. Barron JA, Krizman DB, Ringeisen BR (2005) Laser printing of single cells: statistical analysis, cell viability, and stress. Ann Biomed Eng 33:121–130
22. Koch L, Gruene M, Unger C et al (2013) Laser assisted cell printing. Curr Pharm Biotechnol 14:91–97
23. Kunwar P, Xiong Z, Zhu Y et al (2019) Hybrid laser printing of 3D, multiscale, multimaterial hydrogel structures. Adv Opt Mater 7:1900656. https://doi.org/10.1002/adom.201900656
24. Yeong WY, Chua CK, Leong KF et al (2004) Rapid prototyping in tissue engineering: challenges and potential. Trends Biotechnol 22:643–652
25. Landers R, Hübner U, Schmelzeisen R et al (2002) Rapid prototyping of scaffolds derived from thermoreversible hydrogels and tailored for applications in tissue engineering. Biomaterials 23:4437–4447
26. Peltola SM, Melchels FP, Grijpma DW et al (2008) A review of rapid prototyping techniques for tissue engineering purposes. Ann Med 40:268–280
27. Liu IH, Chang SH, Lin HY (2015) Chitosan-based hydrogel tissue scaffolds made by 3D plotting promotes osteoblast proliferation and mineralization. Biomed Mater 10:035004. https://doi.org/10.1088/1748-6041/10/3/035004
28. Pan T, Song W, Xin H et al (2021) MicroRNA-activated hydrogel scaffold generated by 3D printing accelerates bone regeneration. Bioact Mater 10:1–14
29. Caprioli M, Roppolo I, Chiappone A et al (2021) 3D-printed self-healing hydrogels via digital light processing. Nat Commun 12:2462. https://doi.org/10.1038/s41467-021-22802-z
30. Lu Y, Mapili G, Suhali G et al (2006) A digital micro-mirror device-based system for the microfabrication of complex, spatially patterned tissue engineering scaffolds. J Biomed Mater Res A 77:396–405
31. Melchels FP, Feijen J, Grijpma DW (2010) A review on stereolithography and its applications in biomedical engineering. Biomaterials 31:6121–6130
32. Sultan MT, Lee OJ, Lee JS et al (2022) Three-dimensional digital light-processing bioprinting using silk fibroin-based bio-ink: recent advancements in biomedical applications. Biomedicines 10:3224. https://doi.org/10.3390/biomedicines10123224
33. Kim SH, Yeon YK, Lee JM et al (2018) Precisely printable and biocompatible silk fibroin bioink for digital light processing 3D printing. Nat Commun 9:1620. https://doi.org/10.1038/s41467-018-03759-y
34. Hong H, Seo YB, Kim DY et al (2020) Digital light processing 3D printed silk fibroin hydrogel for cartilage tissue engineering. Biomaterials 232:119679. https://doi.org/10.1016/j.biomaterials.2019.119679
35. Bhusal A, Dogan E, Nguyen HA et al (2021) Multi-material digital light processing bioprinting of hydrogel-based microfluidic chips. Biofabrication. https://doi.org/10.1088/1758-5090/ac2d78
36. Choi KY, Ajiteru O, Hong H et al (2023) A digital light processing 3D-printed artificial skin model and full-thickness wound models using silk fibroin bioink. Acta Biomater 164:159–174
37. Vaupel S, Mau R, Kara S et al (2023) 3D printed and stimulus responsive drug delivery systems based on synthetic polyelectrolyte hydrogels manufactured via digital light processing. J Mater Chem B 11:6547–6559
38. Koffler J, Zhu W, Qu X et al (2019) Biomimetic 3D-printed scaffolds for spinal cord injury repair. Nat Med 25:263–269
39. Xue D, Zhang J, Wang Y et al (2019) Digital light processing-based 3D printing of cell-seeding hydrogel scaffolds with regionally varied stiffness. ACS Biomater Sci Eng 5:4825–4833

40. Jiang G, Li S, Yu K et al (2021) A 3D-printed PRP-GelMA hydrogel promotes osteochondral regeneration through M2 macrophage polarization in a rabbit model. Acta Biomater 128:150–162

41. Li C, Kuss M, Kong Y et al (2021) 3D printed hydrogels with aligned microchannels to guide neural stem cell migration. ACS Biomater Sci Eng 7:690–700

42. Schoonraad SA, Fischenich KM, Eckstein KN et al (2021) Biomimetic and mechanically supportive 3D printed scaffolds for cartilage and osteochondral tissue engineering using photopolymers and digital light processing. Biofabrication. https://doi.org/10.1088/1758-5090/ac23ab

43. Calvert P (2001) Inkjet printing for materials and devices. Chem Mater 13:3299–3305

44. Le HP (1998) Progress and trends in ink-jet printing technology. J Imaging Sci Technol 42:49–62

45. Li J, Chen M, Fan X et al (2016) Recent advances in bioprinting techniques: approaches, applications and future prospects. J Transl Med 14:271. https://doi.org/10.1186/s12967-016-1028-0

46. Nakamura M, Kobayashi A, Takagi F et al (2005) Biocompatible inkjet printing technique for designed seeding of individual living cells. Tissue Eng 11:1658–1666

47. Saunders RE, Gough JE, Derby B (2008) Delivery of human fibroblast cells by piezoelectric drop-on-demand inkjet printing. Biomaterials 29:193–203

48. Wang X, Ao Q, Tian X et al (2016) 3D bioprinting technologies for hard tissue and organ engineering. Materials 9:802. https://doi.org/10.3390/ma9100802

49. Negro A, Cherbuin T, Lutolf MP (2018) 3D inkjet printing of complex, cell-laden hydrogel structures. Sci Rep 8:17099. https://doi.org/10.1038/s41598-018-35504-2

50. Smith PJ, Morrin A (2012) Reactive inkjet printing. J Mater Chem 22:10965

51. Duffy GL, Liang H, Williams RL et al (2021) 3D reactive inkjet printing of poly-ε-lysine/gellan gum hydrogels for potential corneal constructs. Mater Sci Eng C Mater Biol Appl 131:112476. https://doi.org/10.1016/j.msec.2021.112476

52. Teo MY, Stuart L, Aw KC et al (2018) Micro-reactive inkjet printing of three-dimensional hydrogel structures. MRS Adv 3:1575–1581

53. Tibbits S (2014) 4D printing: multi-material shape change. Archit Des 84:116–121

54. Momeni F, Ni J (2020) Laws of 4D printing. Engineering 6:1035–1055

55. Ge Q, Sakhaei A, Lee H et al (2016) Multimaterial 4D printing with tailorable shape memory polymers. Sci Rep 6:31110. https://doi.org/10.1038/srep31110

56. Ge Q, Dunn CK, Qi HJ et al (2014) Active origami by 4D printing. Smart Mater Struct 23:094007. https://doi.org/10.1088/0964-1726/23/9/094007

57. Kim SH, Seo YB, Yeon YK et al (2020) 4D-bioprinted silk hydrogels for tissue engineering. Biomaterials 260:120281. https://doi.org/10.1016/j.biomaterials.2020.120281

58. Lai J, Ye X, Liu J et al (2021) 4D printing of highly printable and shape morphing hydrogels composed of alginate and methylcellulose. Mater Des 205:109699. https://doi.org/10.1016/j.matdes.2021.109699

59. Tamay DG, Dursun Usal T, Alagoz AS et al (2019) 3D and 4D printing of polymers for tissue engineering applications. Front Bioeng Biotechnol 7:164. https://doi.org/10.3389/fbioe.2019.00164

60. Momeni F, Hassani SMM, Liu X et al (2017) A review of 4D printing. Mater Des 122:42–79

61. Koetting MC, Peters JT, Steichen SD et al (2015) Stimulus-responsive hydrogels: theory, modern advances, and applications. Mater Sci Eng R Rep 93:1–49

62. Chang S, Wang S, Liu Z et al (2022) Advances of stimulus-responsive hydrogels for bone defects repair in tissue engineering. Gels 8:389. https://doi.org/10.3390/gels8060389

63. Bajpai AK, Shukla SK, Bhanu S et al (2008) Responsive polymers in controlled drug delivery. Prog Polym Sci 33:1088–1118

64. Xu FJ, Kang ET, Neoh KG (2006) pH- and temperature-responsive hydrogels from crosslinked triblock copolymers prepared via consecutive atom transfer radical polymerizations. Biomaterials 27:2787–2797

65. Han D, Lu Z, Chester SA et al (2018) Micro 3D printing of a temperature-responsive hydrogel using projection micro-stereolithography. Sci Rep 8:1963. https://doi.org/10.1038/s41598-018-20385-2

66. Nizioł M, Paleczny J, Junka A et al (2021) 3D printing of thermoresponsive hydrogel laden with an antimicrobial agent towards wound healing applications. Bioengineering 8:79. https://doi.org/10.3390/bioengineering8060079

67. Wang X, Yu Y, Yang C et al (2022) Dynamically responsive scaffolds from microfluidic 3D printing for skin flap regeneration. Adv Sci 9:e2201155. https://doi.org/10.1002/advs.202201155

68. Abdullah T, Okay O (2023) 4D printing of body temperature-responsive hydrogels based on poly(acrylic acid) with shape-memory and self-healing abilities. ACS Appl Bio Mater 6:703–711

69. Gupta P, Vermani K, Garg S (2002) Hydrogels: from controlled release to pH-responsive drug delivery. Drug Discov Today 7:569–579

70. Choi AH (2022) Biomaterials and bioceramics-part 2: nanocomposites in osseointegration and hard tissue regeneration. In: Choi AH, Ben-Nissan B (eds) Innovative bioceramics in translational medicine I. Springer series in biomaterials science and engineering, vol 17. Springer, Singapore, pp 47–88

71. Garcia C, Gallardo A, López D et al (2018) Smart pH-responsive antimicrobial hydrogel scaffolds prepared by additive manufacturing. ACS Appl Bio Mater 1:1337–1347

72. Wang F, Li L, Zhu X et al (2023) Development of pH-responsive polypills via semi-solid extrusion 3D printing. Bioengineering 10:402. https://doi.org/10.3390/bioengineering10040402

73. Choi Y, Kim C, Kim HS et al (2021) 3D printing of dynamic tissue scaffold by combining self-healing hydrogel and self-healing ferrogel. Colloids Surf B Biointerfaces 208:112108. https://doi.org/10.1016/j.colsurfb.2021.112108

74. Mun CU, Kim HS, Kong M et al (2023) Three-dimensional printing of hyaluronate-based self-healing ferrogel with enhanced stretchability. Colloids Surf B Biointerfaces 221:113004. https://doi.org/10.1016/j.colsurfb.2022.113004

75. Vítková L, Kazantseva N, Musilová L et al (2023) Magneto-responsive hyaluronan hydrogel for hyperthermia and bioprinting: magnetic, rheological properties and biocompatibility. APL Bioeng 7:036113. https://doi.org/10.1063/5.0147181

76. Tran TS, Balu R, Mettu S et al (2022) 4D printing of hydrogels: innovation in material design and emerging smart systems for drug delivery. Pharmaceuticals 15:1282. https://doi.org/10.3390/ph15101282

77. Bender P, Fock J, Hansen MF et al (2018) Influence of clustering on the magnetic properties and hyperthermia performance of iron oxide nanoparticles. Nanotechnology 29:425705

78. Chu H, Yang W, Sun L et al (2020) 4D printing: a review on recent progresses. Micromachines 11:796. https://doi.org/10.3390/mi11090796

79. Mulakkal MC, Trask RS, Ting VP et al (2018) Responsive cellulose-hydrogel composite ink for 4D printing. Mater Des 160:108–118

80. Bashir S, Hina M, Iqbal J et al (2020) Fundamental concepts of hydrogels: synthesis, properties, and their applications. Polymers 12:2702. https://doi.org/10.3390/polym12112702

81. Lv C, Sun XC, Xia H et al (2018) Humidity-responsive actuation of programmable hydrogel microstructures based on 3D printing. Sens Actuators B Chem 259:736–744

82. Hiendlmeier L, Zurita F, Vogel J et al (2023) 4D-printed soft and stretchable self-folding cuff electrodes for small-nerve interfacing. Adv Mater 35:e2210206. https://doi.org/10.1002/adma.202210206

83. Yang GH, Kim W, Kim J et al (2021) A skeleton muscle model using GelMA-based cell-aligned bioink processed with an electric-field assisted 3D/4D bioprinting. Theranostics 11:48–63

84. Li YC, Zhang YS, Akpek A et al (2016) 4D bioprinting: the next-generation technology for biofabrication enabled by stimuli-responsive materials. Biofabrication 9:012001. https://doi.org/10.1088/1758-5090/9/1/012001

85. Finan JD, Guilak F (2010) The effects of osmotic stress on the structure and function of the cell nucleus. J Cell Biochem 109:460–467
86. Mandrycky C, Wang Z, Kim K et al (2016) 3D bioprinting for engineering complex tissues. Biotechnol Adv 34:422–434
87. Malda J, Visser J, Melchels FP et al (2013) 25th anniversary article: engineering hydrogels for biofabrication. Adv Mater 25:5011–5028
88. Ozbolat IT, Yu Y (2013) Bioprinting toward organ fabrication: challenges and future trends. IEEE Trans Biomed Eng 60:691–699
89. Lee HJ, Kim YB, Ahn SH et al (2015) A new approach for fabricating collagen/ECM-based bioinks using preosteoblasts and human adipose stem cells. Adv Healthc Mater 4:1359–1368
90. Groll J, Burdick JA, Cho DW et al (2018) A definition of bioinks and their distinction from biomaterial inks. Biofabrication 11:013001. https://doi.org/10.1088/1758-5090/aaec52
91. Mironov V (2003) Printing technology to produce living tissue. Expert Opin Biol Ther 3:701–704
92. Cohen DL, Lo W, Tsavaris A et al (2011) Increased mixing improves hydrogel homogeneity and quality of three-dimensional printed constructs. Tissue Eng Part C Methods 17:239–248
93. Kačarević ŽP, Rider PM, Alkildani S et al (2018) An introduction to 3D bioprinting: possibilities, challenges and future aspects. Materials 11:2199. https://doi.org/10.3390/ma11112199
94. Santos MI, Reis RL (2010) Vascularization in bone tissue engineering: physiology, current strategies, major hurdles and future challenges. Macromol Biosci 10:12–27
95. Byambaa B, Annabi N, Yue K et al (2017) Bioprinted osteogenic and vasculogenic patterns for engineering 3D bone tissue. Adv Healthc Mater. https://doi.org/10.1002/adhm.201700015
96. Grayson WL, Bunnell BA, Martin E et al (2015) Stromal cells and stem cells in clinical bone regeneration. Nat Rev Endocrinol 11:140–150
97. Tao J, Zhu S, Liao X et al (2022) DLP-based bioprinting of void-forming hydrogels for enhanced stem-cell-mediated bone regeneration. Mater Today Bio 17:100487. https://doi.org/10.1016/j.mtbio.2022.100487
98. Das S, Pati F, Choi YJ et al (2015) Bioprintable, cell-laden silk fibroin-gelatin hydrogel supporting multilineage differentiation of stem cells for fabrication of three-dimensional tissue constructs. Acta Biomater 11:233–246
99. Skardal A, Devarasetty M, Kang HW et al (2016) Bioprinting cellularized constructs using a tissue-specific hydrogel bioink. J Vis Exp. https://doi.org/10.3791/53606.PMID:27166839; PMCID:PMC4941985
100. Wang D, Guo Y, Zhu J et al (2023) Hyaluronic acid methacrylate/pancreatic extracellular matrix as a potential 3D printing bioink for constructing islet organoids. Acta Biomater 165:86–101
101. De Santis MM, Alsafadi HN, Tas S et al (2021) Extracellular-matrix-reinforced bioinks for 3D bioprinting human tissue. Adv Mater 33:e2005476. https://doi.org/10.1002/adma.202005476
102. Zhang M, Yang F, Han D et al (2023) 3D bioprinting of corneal decellularized extracellular matrix: GelMA composite hydrogel for corneal stroma engineering. Int J Bioprint 9:774. https://doi.org/10.18063/ijb.774
103. Wang Z, Lee SJ, Cheng HJ et al (2018) 3D bioprinted functional and contractile cardiac tissue constructs. Acta Biomater 70:48–56
104. Tarassoli SP, Jessop ZM, Al-Sabah A et al (2018) Skin tissue engineering using 3D bioprinting: an evolving research field. J Plast Reconstr Aesthet Surg 71:615–623
105. Shi Y, Xing TL, Zhang HB et al (2018) Tyrosinase-doped bioink for 3D bioprinting of living skin constructs. Biomed Mater 13:035008. https://doi.org/10.1088/1748-605X/aaa5b6
106. Admane P, Gupta AC, Jois P et al (2019) Direct 3D bioprinted full-thickness skin constructs recapitulate regulatory signaling pathways and physiology of human skin. Bioprinting 15:e00051. https://doi.org/10.1016/j.bprint.2019.e00051
107. Kang D, Liu Z, Qian C et al (2023) 3D bioprinting of a gelatin-alginate hydrogel for tissue-engineered hair follicle regeneration. Acta Biomater 165:19–30

108. Singh A, Peppas NA (2014) Hydrogels and scaffolds for immunomodulation. Adv Mater 26:6530–6541
109. Pérez-Herrero E, Lanier OL, Krishnan N et al (2023) Drug delivery methods for cancer immunotherapy. Drug Deliv Transl Res. https://doi.org/10.1007/s13346-023-01405-9
110. Kim D, Jo S, Lee D et al (2023) NK cells encapsulated in micro/macropore-forming hydrogels via 3D bioprinting for tumor immunotherapy. Biomater Res 27:60. https://doi.org/10.1186/s40824-023-00403-9

Chapter 3
Self-healing Hydrogels

Self-healing smart materials have been the pinnacle of regenerative tissue engineering, and in the medical arena, the question of whether we can fabricate synthetic bioprinted and/or bioinspired tissues and scaffolds that can repair themselves if damaged has motivated many researchers in tissue engineering. According to Ghosh, self-healing can be described as the capability of a material to recover or repair (heal) damages it sustained autonomously and automatically without any external intervention. However, the action of self-healing cannot be carried out autonomously and the incorporation of self-healing properties in synthetic materials will require an external trigger [1].

Self-healing engineering materials with characteristics that allow them to heal themselves when damaged by thermal, mechanical, or other mechanisms and restore their original sets of properties include a variety of polymers, metals, ceramics, and composites. The lifespan of any construct or device would be extended significantly through the development and utilization of materials that autonomously repair on the microscopic level in situ prior to suffering macroscopic failures [2]. Typically, self-healing materials are designed to detect, stop, and even reverse damage, without the need for any chemical or physical stimuli applied externally if possible. This notion has been postulated to be the driving force behind the emerging class of self-healing materials that are gifted with the inherent capacity to self-repair when damaged caused by chemical and physical stresses within the intended environment [2].

First-generation self-healing materials are often referred to as extrinsic self-healing and they can repair defects irreversibly but the structure is not restored back to its original form. This is mainly achieved using healing agents encapsulated in tubes or capsules and are released upon breakage [2]. It has been stated that blood clotting and scar formation from a cut or wound follow the concept of first-generation self-healing in which both are aimed at stopping and repairing damage but do not restore the tissue to its undamaged and original state [2]. Second-generation self-healing materials are known as intrinsic self-healing and they are based on the chemistry

© The Author(s), under exclusive license to Springer Nature Singapore Pte Ltd. 2024
A. H. Choi and B. Ben-Nissan, *Hydrogel for Biomedical Applications*, Tissue Repair and Reconstruction, https://doi.org/10.1007/978-981-97-1730-9_3

of dynamic bonds and any interaction or bond that is reversible under equilibrium conditions can be thought of as dynamic. Furthermore, they can be categorized as either covalent or non-covalent [3, 4].

Intrinsic self-healing polymers can be described as materials that are able to repair damages from molecular to macroscale level through an increase in the mobility of polymeric chain locally and temporarily. Typically, such characteristic is centered on certain polymeric molecular structures that permit a noticeable interchain mobility once a small quantity of energy such as UV or temperature is provided. Upon the removal of the stimulus, the physical and chemical bond strength will undergo a restoration process. In comparison to extrinsic self-healing materials, intrinsic self-healing materials offer the possibility of repairing partially or fully the initial properties of the material several times, and the healing process can be applied to both thermoset polymers and thermoplastics as well as elastomers and hydrogels [5].

Self-healing hydrogels are of particular interest due to their ability to repair structural damages and recover their original functions, specifically in tissue engineering, and this healing capacity is derived from either reversible chemical or physical bonds or a combination of both [6]. In addition, self-healing hydrogels with shear-thinning properties can potentially be utilized in bioprinting for the fabrication of tissue engineering constructs.

3.1   Self-healing Mechanism and Chemistry

The so-called mobile phase is one common principle centered on self-healable hydrogels, and theory is that the combination of reconnection and mass transfer of broken links within the hydrogel matrix results in the closure of cracks [7]. In general, either covalent or non-covalent bonds are the driving force behind the reconnection within this matrix.

Self-healing hydrogels can be fabricated through non-covalent and/or dynamic covalent bond interactions. The dynamic equilibrium between dissociation and recombination of various interactions causes the hydrogel to repair damages. In general, dynamic covalent bonds demonstrate stable and slow dynamic equilibriums; on the other hand, non-covalent interactions display rapid and fragile dynamic equilibriums [8].

Chelation can be described as the process of forming multiple coordination bonds between a transition metal ion and organic molecules resulting in the sequestration of the metal [9]. A ligand is an ion or molecule with two or more atoms that can be donated easily to form a covalent bond [10]. Highly complex lattice structures are formed as the transition metal ion is surrounded by the ligands. As a result, the binding energy of the complexes is in general stronger than covalent bonds based on the lattice structure of these metal complexes in addition to the many donor atoms involved. In comparison to covalent bonds, the unique feature of chelation is that they simultaneously demonstrate elasticity, reversibility, and high adhesivity [7].

Subsequently, self-healable hydrogels that are also highly adhesive can be prepared using chelation [11–14].

3.1.1 Dynamic Covalent Bonding

Self-healing hydrogels can be prepared using dynamic covalent chemistry such as Diels–Alder reaction, boronate ester complexation, imine formation, and disulfide exchange [7, 8]. The ability to reconnect without the intervention of any physical stimuli makes dynamic covalent bonds unique and this is completely different to standard covalent bonds. Consequently, dynamic covalent bonds combine the reversibility of non-covalent interactions with the stability of covalent bonds, and the result is a material that possesses healing capacity and can function without any external input [7].

As discussed earlier, the Diels–Alder reaction can prepare covalently crosslinked hydrogels without the use of any metal catalysts and the reaction will remain efficient [15–17]. Numerous features are displayed by the Diels–Alder reaction, and one of the interesting features is that the reaction is thermoreversible, implying that the polymeric material will undergo reversible deposition at elevated temperatures (greater than 100 °C) back to diene and dienophile [17]. According to the study by Chen et al., such a polymer can repeatedly repair itself in mild conditions. In addition, the process is fully reversible, and the fractured area of the polymer could be restored several times without the use of catalysts, addition monomers, or special surface treatment of the fractured interface [18]. Despite its potential as a stimulus-responsive material, its role in tissue engineering and within the biomedical arena is limited due to the high temperature and the time required to perform self-repairing [7, 8]. A study was attempted to prepare a hydrogel based on dextran that could self-heal under physiological conditions. According to the authors, a longitudinal depth of scratch on the hydrogel surface was almost completely healed at 37 °C after 7 h [19]. Other researchers have also investigated the possibility of preparing self-healing hydrogels for potential applications in the tissue engineering and biomedical arena using the Diels–Alder reaction [20–24].

The amalgamation of diol and boronic acid is used to create the reversible boronate ester bond, and conceivable the most vital chemical aspect resulting in the use of boronic acid in biomedical applications is its ability to form reversible covalent complexes with 1,2- or 1,3-diols [7, 8, 25]. However, it is well established that the stability of this reaction is extremely sensitive to pH, and the resultant self-healing efficacy is sensitive to changes in pH as well. It has been suggested that complexation of diols with boronic acid takes place at pH values above the diol pK_a and is postulated to be triggered by diol ionization [26, 27]. In reality, the formation of diol-boronic acid link can only take place if the pH level is the same or exceeds the pKa value of boronic acid, which is in general greater than 8 pKa [7]. Given the fact that most tissues within the human body function at neutral pH and cells perish at pH greater than 8, this could prove to be a disadvantage [7]. Self-healing hydrogels were synthesized in a study based on dynamic covalent bond formation between phenylboronic acid and cis-diol-modified PEG macromonomers. The resultant hydrogel displayed glucose-responsive size-dependent release of proteins and tunable mechanical properties that are responsive to pH. Furthermore, in vitro analysis revealed the hydrogel was cytocompatible and displayed a typical foreign body reaction without chronic inflammation in vivo [28]. Later, a study was conducted to examine the fabrication of boronate ester-crosslinked hydrogels using 2-acrylamidophenylboronic acid and its ability to self-heal under both acidic and neutral pH. The authors suggested the stabilization of crosslinks formed in those conditions was supported by the internal coordination of boronic acid monomer, and such a hydrogel could potentially be used for self-healing applications under physiological pH [29]. First suggested by He et al. [30] and later by Tseng et al. [31], hydrogels composed of PEG diacrylate and dithiotheritol with borax as the catalyst and a glucose-sensitive motif were fabricated. According to the study by He et al., borax is quite efficient as a catalyst and rapid gelation can be achieved at room temperature and under ambient conditions. Moreover, it induces the simultaneous formation of covalent thioether and transient boronate ester bonds when used as a catalyst. They further stated that the obtained hydrogel was pH and thermal responsive, and the hydrogel could be healed within half an hour without external stimulus [30]. On the other hand, the study by Tseng et al. focused on the possibility of utilizing the glucose-sensitive nature of the obtained hydrogel as sacrificial materials for the creation of complex but easily removable structures for building vascularized tissue constructs [31].

Another type of dynamic covalent bond that is used in self-healable hydrogels is the disulfide bond. In essence, they are based on thiol-disulfide exchange reactions utilizing the dynamic properties of disulfide bonds enabling the formation of reversible crosslinks under neutral and alkaline pH conditions in the presence of nucleophilic thiols [4, 7]. However, the thiol groups need to be oxidized, and consequently, the reaction is highly sensitive to the pH value and requires the involvement of an oxidation agent, rendering some of the manufacturing protocols for such hydrogels cytotoxic for cells [7]. In addition, upon aerial oxidation of thiolate groups, thiol-disulfide exchange reactions will lose their reversible and dynamic nature that could hinder the self-healing of hydrogels [4].

The reaction between a nucleophilic group such as amine or hydrazine and the electrophilic carbon of ketones or aldehydes results in the creation of reversible quasicovalent imine or hydrazone bonds. This is referred to as Schiff base reactions [4]. An imine bond can be considered as a strong covalent bond that can occur at both acidic and neutral pH conditions. In this reaction, a water molecule is generated, and if it is not removed, then the reaction can still revert to hydrolysis, and as a result, a dynamic equilibrium is possible under certain scenarios [7]. The reaction is highly selective and has been extensively utilized as dynamic covalent bonds to create self-healing hydrogels by mixing two-component systems containing the reactive motifs [4, 32]. Imine-type reactions are particularly attractive given the fact that they possess tunable rate of hydrolysis at physiological pH and a range of equilibrium contents. Generally, the reaction is chemically specific and orthogonal and has been explored for bioconjugation strategies [33]. The physical properties of a gel can be adjusted using the knowledge of physiological pH and equilibrium contents for various imine linkages. In most cases, the rate of crosslink rearrangement and the viscoelastic response can be adjusted using the physiological pH, and the stiffness of the gel is governed by the equilibrium content [33]. Subsequently, the speed and capacity for a hydrogel to self-heal are determined by the involved chemical and their specific equilibrium content [4, 33]. Using the Schiff-based reaction, a recent study was attempted to fabricate an injectable hydrogel between ε-polylysine-coated manganese dioxide nanosheets and insulin-loaded self-assembled aldehyde Pluronic F127 micelles for possible application in the promotion of multidrug-resistant-infected diabetic wound healing [34]. Other studies were performed to examine the potentials of fabricating injectable hydrogels for regenerative therapies [35–40].

Furthermore, given the fact that an amine group is involved in imine bonds, amino-rich polymers such as chitosan are often used to fabricate self-healable hydrogels by combining them with other aldehyde-functionalized polymer such as hyaluronic acid [4]. A study examined the prospect of fabricating injectable hydrogel composites from quaternized chitosan and benzaldehyde-terminated Pluronic F127 under physiological conditions as wound dressing for joint skin damage [41]. According to the authors, excellent mechanical property and self-healing ability were suggested to be the result of the combined dynamic Schiff base bond and the Pluronic F127 micelle crosslinking within the same hydrogel. Moreover, the obtained hydrogel displayed excellent malleability and compressible property as well as self-healing capacity under multicyclic deformation. As a possible wound dressing, the authors claimed the hydrogels exhibited excellent antibacterial property, biocompatibility, and efficient blood clotting capacity. In vivo examination was also carried out using histopathological investigation, biochemical analysis, immunofluorescence staining, and wound contraction area to determine its therapeutic effects as a hydrogel dressing. Later, a similar study was carried out using quaternized chitosan and oxidized hyaluronic acid as theoretical bases for fabricating wound dressing based on a photothermally improved polysaccharide-based hydrogel that is pH-responsive with antibacterial ability [42]. Likewise, cysteine-modified carboxymethyl chitosan, sodium oxidized alginate, and but-3-yn-2-one were used to prepare hydrogel composite wound dressing via Schiff base and thiol-alkynone double crosslinking. The authors claimed the hydrogel demonstrated adequate in vivo biocompatibility and antibacterial effect, and it is possible to inhibit inflammation and promote skin tissue regeneration in animal models [43].

Derivatives of imine such as oxime and acylhydrazone possess great stability, and they are normally formed from a condensation reaction between a hydrazine and an aldehyde group [44]. They have also been examined for the synthesis of self-healing hydrogels for possible use in regenerative applications [20, 45–48].

3.1.2   Non-covalent Bonding

Non-covalent interactions such as hydrophobic interaction, hydrogel bonding, and electrostatic interaction can also be used in the production of self-healing hydrogels. In comparison to covalent bonding discussed above, non-covalent interactions are more sensitive to environmental conditions such as temperature and pH, and it is less stable [7, 8].

Regarded as one of the most common non-covalent interactions found in nature, hydrogel bonds can still play a role in the mechanical properties of hydrogels despite being several times weaker than covalent and ionic bonds, but only when there are a lot of them present in the hydrogel matrix and only then their contribution is significant. More importantly, the speed in which the association and dissociation of hydrogen bonds occur is the driving force behind the rapid healing time demonstrated by repairable hydrogels. Hydrogel bonds, as a result of their weakness, can be combined with other stronger bonds to produce self-healable hydrogels with the required combination of rapid healing and mechanical properties [7]. Self-healing hydrogels based on hydrogen bonding were often investigated with the addition of other chemical moieties such as ureidopyrimidinone units for applications like tissue regeneration and drug delivery systems [49–53]. Dankers et al. fabricated self-healing hydrogels based on PEG, and according to the authors, this hydrogel based on its nonlinear structural formation allows active proteins to be incorporated to the hydrogel during formation, and once implanted, it releases the protein in vivo through erosion of both the protein and polymer by dissolution [49]. Later, the group developed a pH-switchable supramolecular hydrogel that turned into a liquid at a pH greater than 8.5 and with a viscosity low enough to enable passage through a long catheter while rapidly forming a hydrogel once in contact with tissue. The local in vivo delivery of MRI contrast agents and growth factors reduced the amount of scar collagen in a chronic myocardial infarction animal model [50]. Similarly, the use of a supramolecular hydrogel with a rapid self-integrating property might potentially offer an alternative to the regeneration of tissue complexes if the hydrogel is biocompatible, biodegradable, and capable of releasing biomolecules in a sustained manner. The authors stated the cartilage-bone tissue complex was regenerated successfully in a subcutaneous implantation model using hydrogels with selected cells and biomolecules [51].

Hydrophobic interactions are marginally stronger than hydrogen bonding and it occurs due to the aggregative hydrophobes in aqueous media. Furthermore, the process is easier to control, and they can be modified according to the shape of the hydrophobes as well as the number of hydrophobic moieties on them [7, 8]. Host–guest interaction is a widely exploited non-covalent interaction that can be used to bind two chemical entities together, in which the macrocyclic host moiety is inserted inside the guest moiety to create a unique structure of the inclusion complexation. Moreover, host–guest interaction is a hydrophobic approach that could be used to prepare hydrogels [8]. Studies have shown that supramolecular hydrogels formed using this interaction possess an autonomously self-healing ability but a number of these hydrogels rely on the presence of external stimuli such as light to trigger the self-healing behavior [8, 54–56]. In contrast, hydrogels that can self-heal without the intervention of external stimulus have also been attempted for applications such as wound dressing and tissue regeneration [56–59], and these hydrogels have shown other capacities such as 3D printability [56] and injectability [58–60]. Apart from host–guest interactions, the use of micelle is another hydrophobic scheme that can be used in the preparation of self-healable hydrogels by incorporating surfactants and amphiphilic polymers into the hydrogel structure. The cyclic dissociation and re-association of the micelles contribute to the self-healing mechanisms of such hydrogels [7]. Additionally, a much simpler preparation approach is needed for non-ionic surfactant micelle-laden hydrogels but it remains a challenge to identify appropriate polymer-surfactant combination, especially due to the complex phase behavior of these systems [61]. Self-healing micelle-based hydrogel for the controlled delivery of hydrophobic drugs was investigated in a study where the authors attempted to gain insight into the formulation-structure–property relationship such as the choice of surfactant, ratio between polymer and surfactant, hydrogel structure, and drug transport [61].

Ionic bonding, as an alternative approach, can also be utilized to form self-healing hydrogels via reversible electrostatic interactions between oppositely charged moieties, and such interactions can occur through ionic bridges between same charged polymers mediated by oppositely charged ions or it could happen between oppositely charged polymers [7, 8]. Moreover, self-healing hydrogels can be prepared through reversible electrostatic interactions occurring in polyampholytes [8]. Polyampholyte gels have a water content between 50 and 70% at an equilibrium state, which is much lower than conventional hydrogels. They have a supramolecular structure and possess several characteristics such as high toughness, resistant to fatigue, and are strongly viscoelastic. More importantly, they can 100% self-recover, and some gels displayed partial self-healing after cutting and solvent-induced shape memory effect [62]. Above all, they are non-toxic and antifouling to cell adhesion. Self-healable porous polyampholyte hydrogel was attempted in a study, and according to the authors, the hydrogel displayed high stability in saline solution. The biocompatibility was confirmed using a cytotoxicity test of cells [63]. Later, a polyampholytic terpolymer hydrogel with enhanced skin adhesion was fabricated using an approach that enables the structure of ion-pair association to be adjusted, which also functions as crosslinks in the hydrogel through the addition of an extra neutral monomer component into the network without altering the total charge balance [64]. However, their brittleness and inelasticity are major disadvantages despite their simplicity throughout the fabrication process. Mixing ionically crosslinked hydrogels with a covalently crosslinkable polymer is a method that can be used to address these drawbacks and the product is a double-network hydrogel [7].

3.1.3 Double-Network Hydrogel

The relationship between toughness, strength, and high-water content has always been a challenge during the design and development of hydrogels. Another issue is to discover a mean to improve the strength of the hydrogel while at the same time increasing the self-healing kinetics. In general, the approach used to enhance the mechanical properties of the hydrogel is to increase the number of crosslinks within the matrix, but this also results in a significant increase in the healing time as the mass transfer into the crack site is reduced. It has been suggested that all of these issues can be resolved through the development of double-network hydrogels as they amalgamate a much weaker network typically composed of reversible crosslinks with a rigid and strong network [7].

In 2016, a study was carried out to create a generalizable approach for the preparation of double-network hydrogels that addressed the necessity for rapid self-healing, injectable delivery, and cytocompatibility through tandem supramolecular host–guest assembly and secondarily formed covalent crosslinks [65]. Non-covalent interactions were utilized as a sacrificial network to toughen covalently crosslinked hydrogels based on hyaluronic acid. The authors commented that the shear-thinning of supramolecular bonds permitted the hydrogel injectability subcutaneously and intramyocardially ex vivo as well as into articular cartilage defects, while gentle reaction conditions allowed cell encapsulation with high viability.

Other researchers also investigated the development of injectable self-healing hydrogels that could provide a certain degree of mechanical performance and tissue adhesion for potential utilization as a wound dressing [66–71]. Furthermore, a double-network hydrogel via a "one pot method" was prepared in an effort to incorporate photothermal antibacterial activity into the network. According to the authors, the hydrogel displayed ability to inhibit *Staphylococcus aureus* and *Escherichia coli* in vitro and demonstrated healing when used to treat wounds infected by *S. aureus* by promoting collagen deposition and skin appendage formation [70]. Similarly, another study investigated the potential use of tannic acid as an effective hemostatic and antibacterial component. In addition, the hydrophobic moieties of tannic acid also provided the hydrogel with a water-shielding effect, rendering it non-swellable [71]. Antibacterial effects were tested against *S. aureus* and *E. coli*, while in vivo biocompatibility and hemostatic ability were investigated using a mouse model. In addition to wound dressing, a recent study was attempted to develop an in situ forming bioinspired double-network hydrogel that imitated the microarchitecture of aggrecan and displayed enhanced stiffness at physiological temperature. According to the authors, chondrocytes cultured in the hydrogel displayed excellent cell viability, extended period of cell proliferation, and the production of cartilage-specific matrix. Furthermore, in vivo rabbit cartilage defect model showed evidence of cartilage regeneration using chondrocyte-laden hydrogel [72].

3.2 3D/4D Printing

The development and advancements in 3D-printed self-healing hydrogels will allow researchers to enhance the flexibility and performance of soft tissue engineering scaffolds in a significant manner. Self-healing hydrogels, when fabrications using extrusion-based 3D printing approach, can recover their mechanical integrity and heal autonomously once it is extruded through the nozzle. However, a vital issue that must be considered when fabricating self-healing hydrogels via extrusion-based 3D printing approach is the gel-fluid transition generated by the shear stress during printing and the rapid self-healing ability of the hydrogel as soon as it is printed [73].

According to the studies by Ouyang et al. [74] and later by Bertsch et al. [4], several factors need to be considered during 3D printing of hydrogels using an extrusion-based approach. The primary concern is centered on the relationship between stability of the construct and the printability of the hydrogel, implying that the hydrogel ink must be viscous enough so that it can be extruded from the print head or nozzle through the application of a mechanical force by the printer. As soon as it is printed, the hydrogel must undergo rapid gelation, and more importantly, it must demonstrate sufficient mechanical strength to support the printed structure [74]. The term extrudability refers to the capacity to extrude the hydrogel ink through a nozzle under a reasonable pressure, and for the hydrogel to do so, it must possess sufficiently low viscosity or shear-thinning property [4].

The combination of multiple crosslinking approaches such as double-network hydrogels and/or dual-crosslinked hydrogels is being increasing used in the design of self-healing injectable hydrogels [4]. It has been postulated that a dual-crosslinked hydrogel would enable the production of 3D constructs with prolonged persistence, well-defined structures, and reinforced stiffness in comparison to conventional covalent crosslinked self-healing hydrogels [73]. Furthermore, a study has highlighted the significance of crosslinking on the printing of stable hydrogel structures. A printable hydrogel ink based on a dual-crosslinking hyaluronic acid system that encompassed both shear-thinning and self-healing behavior via guest–host bonding and covalent crosslinking for stabilization using photopolymerization was prepared. Observations from their study revealed long-term stability was not created when either guest–host bonding or covalent crosslinking was used alone due to network relaxation after printing or dispersion of the ink prior to stabilization, respectively. On the other hand, dual-crosslinking approach enabled rapid stabilization of the printed structure via supramolecular bonds immediately upon extrusion that induced short-term stability until covalent crosslinking resulted in further stabilization. Additionally, the author commented that this dual-crosslinking approach did not require the use of other support material during printing and the introduction of a secondary material can be avoided [74].

In general, double-network hydrogels possess interpenetrating polymer network structures that demonstrate a significantly higher mechanical strength than the conventional single-network hydrogel from each component. Typically, one network is soft and ductile and can withstand large deformation, while the other is stiff and brittle and could be fractured to dissipate energy. However, this type of double-network hydrogel lacks elasticity due to the permanent, irreversible rupture of the covalent bond in the stiff and brittle network. Numerous physical networks with reversible non-covalent associations have been introduced into double-network hydrogels for energy dissipation to obtain highly elastic and high-strength hydrogels [75]. In addition, the combination of static covalent crosslinks with dynamic covalent crosslinks and utilizing the reversible dynamic covalent crosslinks as the energy dissipation mechanism has been explored to improve the printability of the hydrogel [76]. A study was attempted to develop a dynamic/photocrosslinking double-network hydrogel bioink based on gelatin and hyaluronic acid. The bioink was prepared by blending amino- and aldehyde-modified hyaluronic acid with gelatin methacrylate, with hydrazone-crosslinked hyaluronic acid as the dynamic network and photocrosslinked gelatin methacrylate as the second static network. According to the authors, the hyaluronic acid network through dynamic hydrazone crosslinking was shear-thinning and self-healing when this double-network hydrogel was applied as bioinks in 3D extrusion printing. The gelatin methacrylate was photocrosslinked after printing to reinforce the entire double-network hydrogel printed scaffold and to improve its mechanical properties [76].

It has been suggested that as a consequence of triple helices and hydrogen bonds, the physical 3D network of gelatin hydrogel would collapse suddenly at temperatures over 30–35 °C, which is under the physiological temperatures. This also restricts gelatin from withstanding temperature used during extrusion-based 3D printing, as it cannot be heated to control the viscosity [77]. Hence, the fabrication of a gelatin hydrogel that could be processed thermally and adapted for extrusion-based 3D printing would be beneficial in the fabrication of tissue engineering scaffolds. In the study by Heidarian and Kouzani, a self-healing nanocomposite double-network gelatin hydrogel was examined in which gelatin was first blended with carboxyl methyl chitosan to increase its amino groups' content. The 3D structure of the gelatin hydrogel was then formed via an in situ formation of dynamic imine bonds from the double network of gelatin/carboxyl methyl chitosan in the presence of dialdehyde-functionalized bacterial nanocellulose, which also functioned as a nanofiber crosslinker that can simultaneously crosslink and reinforce the double network through the formation of dynamic 3D imine bonds [77]. According to the authors, the hydrogel displayed high thermal stability during extrusion-based 3D printing using a temperature above 37 °C, and it could self-heal in a neutral and physiological environment without the assistance of any external stimuli or additives [77].

In addition to extrusion-based approach, a study has postulated that self-healing hydrogels could also be 3D printing using a commercially available digital light processing printer. The hydrogel was prepared by combining a covalent network of acrylic acid and PEG diacrylate with a physical network of poly(vinyl alcohol). According to the authors, the hydrogel can self-heal rapidly at room temperature without the need for external stimuli [78].

3.2.1 Bioprinting

As discussed above, one of the key design considerations when utilizing extrusion-based 3D printing is that the hydrogel ink needs to be of sufficiently low viscosity to allow extrusion and mechanically sufficient to enable the printed structure to maintain its shape over time (with and without cell inclusion). The viability of cells is often improved in hydrogels that are loosely crosslinked as it allows the diffusion of wastes and nutrients while hydrogels with high viscosity will allow it to maintain its shape but may cause cells to experience damaging shear stresses during the extrusion process [79]. During extrusion, shear-thinning (injectable) hydrogels will fluidize due to the application of mechanical forces, and subsequently, shear stress might be one of the primary factors for cellular damage induced during the bioprinting process [75]. Cell damage induced by shear stress could be due to variables such as increasing the ink viscosity, the speed of extrusion, or reducing the nozzle diameter. In contrast, shape fidelity can be improved by using an ink that has high viscosity and the printing resolution can be enhanced using small diameter nozzles during printing. Accordingly, a balance must be reached between cell viability and hydrogel ink printability [75]. Several studies have been carried out to examine the relationship between hydrogel ink properties and printability using extrusion-based printers to achieve both adequate printability and high cell viability [79–82].

As a unique dynamic covalent bond, the hydrazone bond is formed between an electron-deficient electrophilic aldehyde and an electron-dense nucleophilic hydrazide. It has been suggested that hydrazide crosslinked hydrogel, as a result of its injectability and shear-thinning characteristics, is capable of protecting cells during extrusion-based printing [83]. Using this approach, a hyaluronic acid hydrogel was prepared in a study to investigate the correlation between substrate stiffness and cell viability [79]. During the extrusion printing of the hydrogel, observations reinforced the fact that the magnitude of force applied during extrusion directly impacts cell viability, and the extrusion forces for printing were dependent on printing variables such as hydrogel concentration and the needle gauge used (as nozzle). The authors further suggested dynamic covalent bonds also contributed in protecting the encapsulated fibroblasts from experiencing shear during the extrusion process.

3.2.2 4D Printing of Self-healing Hydrogels with Shape Memory

As previously mentioned, smart hydrogels are one of the two most widely used polymeric materials for 4D printing. In recent years, only a few studies have been carried out to examine the feasibility of printing hydrogels that can concurrently possess self-healing and shape memory abilities [84–86]. This was motivated by the fact that shape memory hydrogels with a single function can no longer fulfill the requirements for wider potential applications. In addition, using one interaction for the shape memory effect and another for self-healing, self-healing process during shape memory performance or shape memory behavior after self-healing process can also be achieved [87].

Recently, the study by Abdullah and Okay described the fabrication of 4D-printed hydrogels based on hydrophilic poly(acrylic acid) chains containing different molar fractions of hydrophobic C16A segments in the presence of a photoinitiator via the stereolithography technique using a commercial resin printer [85]. The authors claimed shape memory effect was observed on the printed hydrogel near the human body temperature and this was achieved by adjusting the C16A monomer content, which in turn affected the melting and crystallization temperature of the hydrogel. In addition, structural and physical damage could be recovered by heating the hydrogel above the melting temperature of the C16A crystalline domains.

Utilizing a dynamic covalent imine/Diels–Alder network, multiresponsive hydrogels consisted of modified gelatin and PEG-based polymers were attempted [85]. According to the authors, the hydrogel, through further secondary crosslinking with a hyperbranched triethoxysilane reagent that contained multiple supramolecular hydrogel bonding, displayed both self-healing and temperature-responsive shape memory effect [86].

Aside from shape memory and self-healing abilities, the study by Wu and Hsu also investigated whether cryopreserving property can also be incorporated into the hydrogel system consisting of photo-/thermos-responsive gelatin-based biomaterials and biodegradable polyurethane nanoparticles [84]. In their study, the self-healing property of the hydrogel was postulated to be associated with the formation of reversible ionomeric interaction between NH_3^+ group on the gelatin chains and the COO group of polyurethane nanoparticles. They further commented that the shape-memorizable 4D-printed constructs revealed good shape fixity and shape recovery through the elasticity as well as forming and collapsing of water lattice in the hydrogel. More importantly, they claimed the hydrogel and the printing process could support the continuous proliferation of neural stem cells for up to 14 days, and the individually bioprinted neural stem cells and mesenchymal stem cells in the adjacent, self-healed filaments showed mutual migration and such interaction promoted the cell differentiation behavior. The cryopreserved 4D bioprinted hydrogel at temperatures of -20 or $-80\,°C$ after awakening and shape recovery at $37\,°C$ demonstrated cell proliferation similar to that of the non-cryopreserved control.

3.3 Concluding Remarks

For tissue regeneration applications, creating an ideal environment that could provide cells to proliferate and subsequent tissue regeneration is a vital criterion during the design and fabrication of a biomaterial such as hydrogel. Self-healing hydrogels with shear-thinning properties (injectability) are at the forefront of numerous approaches for regenerative medicine and tissue engineering. Injectable self-healing hydrogels are especially beneficial for utilization at sites which are dangerous or difficult to access such as the central nervous system in which they can be applied to bridge spinal cord lesions. Based on their mechanical characteristics, self-healing hydrogels can be considered as soft hydrogels in the biomedical arena. One of the primary advantages is their shear-thinning characteristic which allows them to be injected into a target site and self-heal and recover their mechanical properties after injection to ensure in situ confinement of the hydrogel.

The use of 3D-printed hydrogel-based scaffolds for tissue repair and regeneration is appealing as they provide cellular environments that are comparable to those found within the human body on top of characteristics such as mechanical properties and high-water content that are similar to the natural extracellular matrix. Additionally, self-healing injectable hydrogels have been explored as printing inks for 3D/4D printing and 3D bioprinting based on the capacity for the hydrogel to fluidize and flow under shear stress followed by rapid self-recovery and self-healing. Despite being one of the fastest evolving areas in tissue engineering, the absence of clear rheological requirements that a hydrogel should possess in order for it to be considered as 3D printing ink is having a negative impact on the development of hydrogels as 3D printing (bio)inks. Subsequently, the design of self-healing hydrogels as printer ink often depends on time-consuming trial-and-error approach to find the optimal combinations of shape fidelity, printability, and in the case of bioinks, cell viability.

The viability of cells is often improved in hydrogels that are loosely crosslinked as it allows the diffusion of wastes and nutrients while hydrogels with high viscosity will allow it to maintain its shape but may result in cells to experience damaging shear stresses during the extrusion process. In contrast, shape fidelity can be improved by using an ink that has high viscosity, and the printing resolution can be enhanced using small diameter nozzles during printing. Cell damage induced by shear stress could be due to variables such as increasing the ink viscosity, the speed of extrusion, or reducing the nozzle diameter.

Given that all the prerequisites for practical application are often not entirely fulfilled, it is somewhat promising to develop shape memory hydrogels with other functionalities such as self-healing. Using one interaction for the shape memory effect and another for self-healing, self-healing process during shape memory performance or shape memory behavior after self-healing process can also be achieved. More research needs to be carried out to explore the 4D printing of hydrogels that can simultaneously possess shape memory and self-healing abilities near physiological temperature.

References

1. Ghosh SK (2008) Self-healing materials: fundamentals, design strategies, and applications. In: Ghosh SK (ed) Self-healing materials. Wiley, New Jersey, pp 1–28
2. Brochu AB, Craig SL, Reichert WM (2011) Self-healing biomaterials. J Biomed Mater Res A 96:492–506

3. Utrera-Barrios S, Verdejo R, López-Manchado MA et al (2020) Evolution of self-healing elastomers, from extrinsic to combined intrinsic mechanisms: a review. Mater Horiz 7:2882–2902
4. Bertsch P, Diba M, Mooney DJ et al (2023) Self-healing injectable hydrogels for tissue regeneration. Chem Rev 123:834–873
5. Garcia SJ (2014) Effect of polymer architecture on the intrinsic self-healing character of polymers. Eur Polym J 53:118–125
6. Rumon MMH, Akib AA, Sultana F et al (2022) Self-healing hydrogels: development, biomedical applications, and challenges. Polymers 14:4539. https://doi.org/10.3390/polym1 4214539
7. Talebian S, Mehrali M, Taebnia N et al (2019) Self-healing hydrogels: the next paradigm shift in tissue engineering? Adv Sci 6:1801664. https://doi.org/10.1002/advs.201801664
8. Liu Y, Hsu SH (2018) Synthesis and biomedical applications of self-healing hydrogels. Front Chem 6:449. https://doi.org/10.3389/fchem.2018.00449
9. van Lith R, Ameer GA (2016) Antioxidant polymers as biomaterial. In: Dziubla T, Allan Butterfield D (eds) Oxidative stress and biomaterials. Academic Press, Massachusetts, pp 251–296
10. Gulcin İ, Alwasel SH (2022) Metal ions, metal chelators and metal chelating assay as antioxidant method. Processes 10:132. https://doi.org/10.3390/pr10010132
11. Krogsgaard M, Behrens MA, Pedersen JS et al (2013) Self-healing mussel-inspired multi-pH-responsive hydrogels. Biomacromolecules 14:297–301
12. Li Q, Barrett DG, Messersmith PB et al (2016) Controlling hydrogel mechanics via bio-inspired polymer-nanoparticle bond dynamics. ACS Nano 10:1317–1324
13. Xiong J, Yang ZR, Lv N et al (2022) Self-adhesive hyaluronic acid/antimicrobial peptide composite hydrogel with antioxidant capability and photothermal activity for infected wound healing. Macromol Rapid Commun 43:e2200176. https://doi.org/10.1002/marc.202200176
14. Yang Y, Shi K, Yu K et al (2022) Degradable hydrogel adhesives with enhanced tissue adhesion, superior self-healing, cytocompatibility, and antibacterial property. Adv Healthc Mater 11:e2101504. https://doi.org/10.1002/adhm.202101504
15. Diels O, Alder K (2010) Synthesen in der hydroaromatischen Reihe. Eur J Org Chem 460:98–122
16. Gregoritza M, Brandl FP (2015) The Diels–Alder reaction: a powerful tool for the design of drug delivery systems and biomaterials. Eur J Pharm Biopharm 97B:438–453
17. Morozova SM (2023) Recent advances in hydrogels via Diels–Alder crosslinking: design and applications. Gels 9:102. https://doi.org/10.3390/gels9020102
18. Chen X, Dam MA, Ono K et al (2002) A thermally re-mendable cross-linked polymeric material. Science 295:1698–1702
19. Wei Z, Yang JH, Du XJ et al (2013) Dextran-based self-healing hydrogels formed by reversible Diels–Alder reaction under physiological conditions. Macromol Rapid Commun 34:1464–1470
20. Yu F, Cao X, Du J et al (2015) Multifunctional hydrogel with good structure integrity, self-healing, and tissue-adhesive property formed by combining Diels–Alder click reaction and acylhydrazone bond. ACS Appl Mater Interfaces 7:24023–24031
21. Zhao J, Xu R, Luo G et al (2016) A self-healing, re-moldable and biocompatible crosslinked polysiloxane elastomer. J Mater Chem B 4:982–989
22. Banerjee SL, Bhattacharya K, Samanta S et al (2018) Self-healable antifouling zwitterionic hydrogel based on synergistic phototriggered dynamic disulfide metathesis reaction and ionic interaction. ACS Appl Mater Interfaces 10:27391–27406
23. Li S, Yi J, Yu X et al (2018) Preparation and characterization of acid resistant double cross-linked hydrogel for potential biomedical applications. ACS Biomater Sci Eng 4:872–883
24. Li DQ, Wang SY, Meng YJ et al (2021) Fabrication of self-healing pectin/chitosan hybrid hydrogel via Diels–Alder reactions for drug delivery with high swelling property, pH-responsiveness, and cytocompatibility. Carbohydr Polym 268:118244. https://doi.org/10.1016/j.carbpol.2021.118244

25. Cambre JN, Sumerlin BS (2011) Biomedical applications of boronic acid polymers. Polymer 52:4631–4643
26. Yan J, Springsteen G, Deeter S et al (2004) The relationship among pKa, pH, and binding constants in the interactions between boronic acids and diols—it is not as simple as it appears. Tetrahedron 60:11205–11209
27. He L, Fullenkamp DE, Rivera JG et al (2011) pH responsive self-healing hydrogels formed by boronate-catechol complexation. Chem Commun 47:7497–7499
28. Yesilyurt V, Webber MJ, Appel EA et al (2016) Injectable self-healing glucose-responsive hydrogels with pH-regulated mechanical properties. Adv Mater 28:86–91
29. Deng CC, Brooks WLA, Abboud KA et al (2015) Boronic acid-based hydrogels undergo self-healing at neutral and acidic pH. ACS Macro Lett 4:220–224
30. He L, Szopinski D, Wu Y et al (2015) Toward self-healing hydrogels using one-pot thiolene click and borax-diol chemistry. ACS Macro Lett 4:673–678
31. Tseng TC, Hsieh FY, Theato P et al (2017) Glucose-sensitive self-healing hydrogel as sacrificial materials to fabricate vascularized constructs. Biomaterials 133:20–28
32. Mo C, Xiang L, Chen Y (2021) Advances in injectable and self-healing polysaccharide hydrogel based on the Schiff base reaction. Macromol Rapid Commun 42:e2100025. https://doi.org/10.1002/marc.202100025
33. Hafeez S, Ooi HW, Morgan FLC et al (2018) Viscoelastic oxidized alginates with reversible imine type crosslinks: self-healing, injectable, and bioprintable hydrogels. Gels 4:85. https://doi.org/10.3390/gels4040085
34. Wang S, Zheng H, Zhou L et al (2020) Nanoenzyme-reinforced injectable hydrogel for healing diabetic wounds infected with multidrug resistant bacteria. Nano Lett 20:5149–5158
35. Chen H, Cheng J, Ran L et al (2018) An injectable self-healing hydrogel with adhesive and antibacterial properties effectively promotes wound healing. Carbohydr Polym 201:522–531
36. Han X, Meng X, Wu Z et al (2018) Dynamic imine bond cross-linked self-healing thermosensitive hydrogels for sustained anticancer therapy via intratumoral injection. Mater Sci Eng C Mater Biol Appl 93:1064–1072
37. Wei Z, Gerecht S (2018) A self-healing hydrogel as an injectable instructive carrier for cellular morphogenesis. Biomaterials 185:86–96
38. Basu S, Pacelli S, Paul A (2020) Self-healing DNA-based injectable hydrogels with reversible covalent linkages for controlled drug delivery. Acta Biomater 105:159–169
39. Li R, Zhou C, Chen J et al (2022) Synergistic osteogenic and angiogenic effects of KP and QK peptides incorporated with an injectable and self-healing hydrogel for efficient bone regeneration. Bioact Mater 18:267–283
40. Jiang H, Xiao Y, Huang H et al (2023) An injectable, adhesive, and self-healing hydrogel with inherently antibacterial property for wound dressing. Macromol Biosci. https://doi.org/10.1002/mabi.202300282
41. Qu J, Zhao X, Liang Y et al (2018) Antibacterial adhesive injectable hydrogels with rapid self-healing, extensibility and compressibility as wound dressing for joints skin wound healing. Biomaterials 183:185–199
42. Xue C, Xu X, Zhang L et al (2022) Self-healing/pH-responsive/inherently antibacterial polysaccharide-based hydrogel for a photothermal strengthened wound dressing. Colloids Surf B Biointerfaces 218:112738. https://doi.org/10.1016/j.colsurfb.2022.112738
43. Zhang Z, Bu J, Li B et al (2022) Dynamic double cross-linked self-healing polysaccharide hydrogel wound dressing based on Schiff base and thiol-alkynone reactions. Int J Mol Sci 23:13817. https://doi.org/10.3390/ijms232213817
44. Belowich ME, Stoddart JF (2012) Dynamic imine chemistry. Chem Soc Rev 41:2003–2024
45. Lü S, Bai X, Liu H et al (2017) An injectable and self-healing hydrogel with covalent cross-linking in vivo for cranial bone repair. J Mater Chem B 5:3739–3748
46. Bo Y, Zhang L, Wang Z et al (2021) Antibacterial hydrogel with self-healing property for wound-healing applications. ACS Biomater Sci Eng 7:5135–5143
47. Li S, Dong Q, Peng X et al (2022) Self-healing hyaluronic acid nanocomposite hydrogels with platelet-rich plasma impregnated for skin regeneration. ACS Nano 16:11346–11359

48. Chen D, Liu X, Qi Y et al (2022) Poly(aspartic acid) based self-healing hydrogel with blood coagulation characteristic for rapid hemostasis and wound healing applications. Colloids Surf B Biointerfaces 214:112430. https://doi.org/10.1016/j.colsurfb.2022.112430

49. Dankers PY, Hermans TM, Baughman TW et al (2012) Hierarchical formation of supramolecular transient networks in water: a modular injectable delivery system. Adv Mater 24:2703–2709

50. Bastings MM, Koudstaal S, Kieltyka RE et al (2014) A fast pH-switchable and self-healing supramolecular hydrogel carrier for guided, local catheter injection in the infarcted myocardium. Adv Healthc Mater 3:70–78

51. Hou S, Wang X, Park S et al (2015) Rapid self-integrating, injectable hydrogel for tissue complex regeneration. Adv Healthc Mater 4:1491–1495, 1423

52. Zhang G, Lv L, Deng Y et al (2017) Self-healing gelatin hydrogels cross-linked by combining multiple hydrogen bonding and ionic coordination. Macromol Rapid Commun. https://doi.org/10.1002/marc.201700018

53. Xu Y, Yang H, Zhu H et al (2020) Self-healing gelatin-based shape memory hydrogels via quadruple hydrogen bonding and coordination crosslinking for controlled delivery of 5-fluorouracil. J Biomater Sci Polym Ed 31:712–728

54. Loh XJ (2014) Supramolecular host–guest polymeric materials for biomedical applications. Mater Horiz 1:185–195

55. Yang Q, Wang P, Zhao C et al (2017) Light-switchable self-healing hydrogel based on host-guest macro-crosslinking. Macromol Rapid Commun. https://doi.org/10.1002/marc.201600741

56. Wang Z, An G, Zhu Y et al (2019) 3D-printable self-healing and mechanically reinforced hydrogels with host-guest non-covalent interactions integrated into covalently linked networks. Mater Horiz 6:733–742

57. Yang M, Tian J, Zhang K et al (2023) Bioinspired adhesive antibacterial hydrogel with self-healing and on-demand removability for enhanced full-thickness skin wound repair. Biomacromolecules. https://doi.org/10.1021/acs.biomac.3c00576

58. Li P, Zong H, Li G et al (2023) Building a poly(amino acid)/chitosan-based self-healing hydrogel via host-guest interaction for cartilage regeneration. ACS Biomater Sci Eng 9:4855–4866

59. Yilmaz-Aykut D, Torkay G, Kasgoz A et al (2023) Injectable and self-healing dual crosslinked gelatin/kappa-carrageenan methacryloyl hybrid hydrogels via host-guest supramolecular interaction for wound healing. J Biomed Mater Res B Appl Biomater 111:1921–1937

60. Jiang X, Zeng F, Yang X et al (2022) Injectable self-healing cellulose hydrogel based on host-guest interactions and acylhydrazone bonds for sustained cancer therapy. Acta Biomater 141:102–113

61. Munoz SZ, Zhadan R, Acosta E (2017) Design of nonionic micelle-laden polysaccharide hydrogels for controlled delivery of hydrophobic drugs. Int J Pharm 526:455–465

62. Sun TL, Kurokawa T, Kuroda S et al (2013) Physical hydrogels composed of polyampholytes demonstrate high toughness and viscoelasticity. Nat Mater 12:932–937

63. Liu JH, Hung YH, Chang KT et al (2020) Self-healable porous polyampholyte hydrogels with higher water content as cell culture scaffolds for tissue engineering applications. ACS Appl Bio Mater 3:5446–5453

64. Lee JH, Lee DS, Jung YC et al (2021) Development of a tough, self-healing polyampholyte terpolymer hydrogel patch with enhanced skin adhesion via tuning the density and strength of ion-pair associations. ACS Appl Mater Interfaces 13:8889–8900

65. Rodell CB, Dusaj NN, Highley CB et al (2016) Injectable and cytocompatible tough double-network hydrogels through tandem supramolecular and covalent crosslinking. Adv Mater 28:8419–8424

66. Chen K, Feng Y, Zhang Y et al (2019) Entanglement-driven adhesion, self-healing, and high stretchability of double-network PEG-based hydrogels. ACS Appl Mater Interfaces 11:36458–36468

67. Yang B, Song J, Jiang Y et al (2020) Injectable adhesive self-healing multicross-linked double-network hydrogel facilitates full-thickness skin wound healing. ACS Appl Mater Interfaces 12:57782–57797

68. Chen C, Zhou P, Huang C et al (2021) Photothermal-promoted multi-functional dual network polysaccharide hydrogel adhesive for infected and susceptible wound healing. Carbohydr Polym 273:118557. https://doi.org/10.1016/j.carbpol.2021.118557

69. Zhou Z, Mei X, Hu K et al (2023) Nanohybrid double network hydrogels based on a platinum nanozyme composite for antimicrobial and diabetic wound healing. ACS Appl Mater Interfaces 15:17612–17626

70. Ren Y, Huang T, Zhao X (2023) Double network hydrogel based on curdlan and flaxseed gum with photothermal antibacterial properties for accelerating infectious wound healing. Int J Biol Macromol 242:124715. https://doi.org/10.1016/j.ijbiomac.2023.124715

71. Park J, Kim TY, Kim Y et al (2023) A mechanically resilient and tissue-conformable hydrogel with hemostatic and antibacterial capabilities for wound care. Adv Sci 13:e2303651. https://doi.org/10.1002/advs.202303651

72. Thomas J, Chopra V, Rajput S et al (2023) Post-implantation stiffening by a bioinspired, double-network, self-healing hydrogel facilitates minimally invasive cell delivery for cartilage regeneration. Biomacromolecules 24:3313–3326

73. Roh HH, Kim HS, Kim C et al (2021) 3D printing of polysaccharide-based self-healing hydrogel reinforced with alginate for secondary cross-linking. Biomedicines 9:1224. https://doi.org/10.3390/biomedicines9091224

74. Ouyang L, Highley CB, Rodell CB et al (2016) 3D printing of shear-thinning hyaluronic acid hydrogels with secondary cross-linking. ACS Biomater Sci Eng 2:1743–1751

75. Xu C, Dai G, Hong Y (2019) Recent advances in high-strength and elastic hydrogels for 3D printing in biomedical applications. Acta Biomater 95:50–59

76. Wang Y, Chen Y, Zheng J et al (2022) Three-dimensional printing self-healing dynamic/photocrosslinking gelatin-hyaluronic acid double-network hydrogel for tissue engineering. ACS Omega 7:12076–12088

77. Heidarian P, Kouzani AZ (2023) A self-healing nanocomposite double network bacterial nanocellulose/gelatin hydrogel for three dimensional printing. Carbohydr Polym 313:120879. https://doi.org/10.1016/j.carbpol.2023.120879

78. Caprioli M, Roppolo I, Chiappone A et al (2021) 3D-printed self-healing hydrogels via digital light processing. Nat Commun 12:2462. https://doi.org/10.1038/s41467-021-22802-z

79. Wang LL, Highley CB, Yeh YC et al (2018) Three-dimensional extrusion bioprinting of single- and double-network hydrogels containing dynamic covalent crosslinks. J Biomed Mater Res A 106:865–875

80. Chung JHY, Naficy S, Yue Z et al (2013) Bio-ink properties and printability for extrusion printing living cells. Biomater Sci 1:763–773

81. Zhao Y, Li Y, Mao S et al (2015) The influence of printing parameters on cell survival rate and printability in microextrusion-based 3D cell printing technology. Biofabrication 7:045002. https://doi.org/10.1088/1758-5090/7/4/045002

82. Kim SW, Kim DY, Roh HH et al (2019) Three-dimensional bioprinting of cell-laden constructs using polysaccharide-based self-healing hydrogels. Biomacromolecules 20:1860–1866

83. Wang H, Zhu D, Paul A et al (2017) Covalently adaptable elastin-like protein-hyaluronic acid (ELP-HA) hybrid hydrogels with secondary thermoresponsive crosslinking for injectable stem cell delivery. Adv Funct Mater 27:1605609. https://doi.org/10.1002/adfm.201605609

84. Wu SD, Hsu SH (2021) 4D bioprintable self-healing hydrogel with shape memory and cryopreserving properties. Biofabrication. https://doi.org/10.1088/1758-5090/ac2789

85. Abdullah T, Okay O (2023) 4D printing of body temperature-responsive hydrogels based on poly(acrylic acid) with shape-memory and self-healing abilities. ACS Appl Bio Mater 6:703–711

86. Wang Z, Gu J, Zhang D et al (2023) Structurally dynamic gelatin-based hydrogels with self-healing, shape memory, and cytocompatible properties for 4D printing. Biomacromolecules 24:109–117

87. Shang JJ, Le XX, Zhang JW et al (2019) Trends in polymeric shape memory hydrogels and hydrogel actuators. Polym Chem 10:1036–1055

Chapter 4
Future of Drug Delivery: Microrobotics and Self-powered Devices

4.1 Microrobots, Micromotors, and Nanomotors

According to the review by Soto et al., micro- and nanomotors (sometimes also referred to as microswimmers, micro- or nanoengines) are often referred to by the robotics community as the first generation of small-scale robots. They are basically small-scale structures that can achieve actuation or locomotion through the conversion of different energy sources [1]. On the other hand, micro- and nanorobots are in essence small-scale structures that are capable of performing preprogrammed tasks through mechanical actuation (Fig. 4.1).

4.1.1 Micromotors and Nanomotors

Synthetic micro- and nanomotors are microscopic or nanoscopic devices that can self-propel by converting the supplied fuel into mechanical motion [2]. Recent advances into the research and development of micro- and nanomotors have made them promising tools in addressing numerous biomedical challenges due to their unique characteristics such as high towing force, cargo loading, and fast motion [3]. Additionally, it has been hypothesized that these synthetic motors can be designed at the molecular level to permit quicker interactions with the biological species. This is of particular significance for biomedical applications based on the reason that higher cell uptake and quicker binding of these motors could potentially result in stronger therapeutic effects to the diseased cells when used as drug delivery systems or earlier detection in cancer therapy [3].

Deposition and assembly are two different approaches that have been explored for the fabrications of micro- and nanomotors. As a versatile methodology, deposition (which is classified as a non-assembling approach) offers the possibility for

© The Author(s), under exclusive license to Springer Nature Singapore Pte Ltd. 2024
A. H. Choi and B. Ben-Nissan, *Hydrogel for Biomedical Applications*, Tissue Repair and Reconstruction, https://doi.org/10.1007/978-981-97-1730-9_4

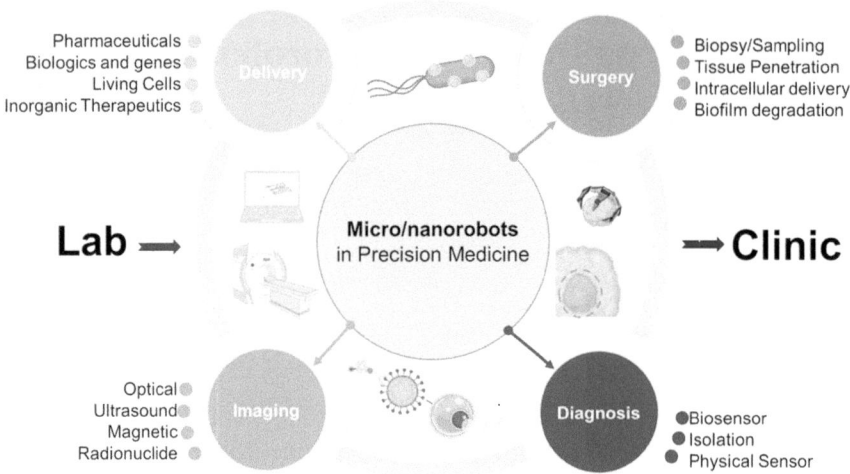

Fig. 4.1 Current trends of micro/nanorobotics in precision medicine, including delivery, surgery, diagnosis, and medical imaging applications. Reprint with permission from [1]. http://creativec ommons.org/licenses/by/4.0/

fabricating synthetic motors with different sizes and shapes as well as the flexibility of using a range of materials. There are two main categories in deposition and both groups can be used. During (electro)chemical deposition, a variety of scaffolds ranging from organic to inorganic materials including hydrogels are suitable via a redox reaction or through an external electric field [2, 4, 5]. In addition, this approach permits the fabrication of synthetic motors based on the porous membrane template, which is typically employed as a cathode for the deposition of materials. By adjusting the fabrication factors such as the deposition time, applied current density, and the pore size and shape of the membrane template, the geometry of the synthetic motors can be governed during deposition [2, 4]. Furthermore, chemical deposition can also be used to prepare hydrogel-based Janus spherical micromotors [6]. It has been hypothesized that synthetic motors with better efficiency for biological applications can be designed using the chemical deposition approach in conjunction with the application of external fields [4]. According to Peng et al., conductive beads were used as templates instead of static membranes and were placed between two electrodes, which can then undergo polarized electrochemical deposition at one hemisphere of the bead [2].

In addition to chemical deposition, physical vapor deposition is another commonly used approach for the preparation of synthetic motors. The process involves the use of ionized argon gas or an electron beam to vaporize the target material followed by the formation of a thin film or coating on the surface of a substrate. Partial shielding may be needed to ensure asymmetry and the anisotropic accumulation of products, which in turn leads to the asymmetry of spherical motors and is the main factor for directional motion [2, 4]. A study was attempted to fabricate micromotors using the vapor deposition approach. Hydrogel micromotors were attained via complex water-in-oil-in-water double emulsion drops and oil-in-water emulsion drops from glass capillary microfluidics and subsequent photopolymerization. In their study, three hydrogel micromotors were investigated: microcapsules with catalyst-containing liquid cores, hydrogel Janus capsules with e-beam evaporated platinum catalyst, and homogeneous Janus particles with e-beam evaporated platinum catalyst [7].

Fabrication based on assembly offers an effective and simple way to integrate various miniature and multicomponent elements such as lipids and inorganic nanoparticles [2, 4]. The fabrication of micro- and nanomotors can be carried out using self-assembly of macromolecules or layer-by-layer (LBL) assembly. Briefly, self-assembly approach involves the spontaneous reorganization of molecules into structures and patterns, while the LBL assembly offers an inexpensive and easy process for multilayer formation and permits the incorporation of different materials within a structure. Hence, the LBL approach can be considered as a bottom-up fabrication technique [4, 8]. Furthermore, the bottom-up assembly approach also provides a simple and versatile means to achieve multifunctionality by enabling the integration of different components into the synthetic motors [9]. In addition, large-scale production of micro- and nanomotors with well-defined structures can be achieved using assembly techniques. However, synthetic motors prepared using the assembly approach are extremely sensitive to changes in the surrounding environment such as biological, chemical, and physical stimuli since they are typically fabricated from soft materials such as surfactants and polymers and are held together by non-covalent interactions [9]. On the other hand, this could be beneficial for applications such as drug delivery. Recently, a study hypothesized the encapsulation of micromotors within a hydrogel "capsule" would be able to protect it from rupture or passivated under the complex biological environment. The micromotors were prepared through layer-by-layer assembly technology and loaded into the Schiff base hydrogel. The asymmetric decomposition of hydrogen peroxide catalyzed by the locally distributed platinum nanoparticles enabled efficient propulsion of the micromotors in low concentration of hydrogen peroxide. The authors postulate such as system would allow for the release of micromotors in response to the environment, reducing external influence on micromotors and improving the sustained-release effect [10].

Although considerable efforts have been made into the fabrication and applications of micro- and nanomotors, hydrogen peroxide is still one of the most used fuel molecules to power catalytic micromotors [4, 11]. It has been widely established that hydrogen peroxide is regarded as a cytotoxic agent whose levels must be kept to a minimal through the action of antioxidant defense enzymes [12]. Additionally, it is considered to induce oxidative stress to cells and results in denaturation of protein. Consequently, utilizing a biocompatible fuel at biologically relevant concentrations or locally available chemicals as fuel sources for micro- and nanomotors offers an alternative solution to the issue of biocompatibility as well as for successful in vivo application [4].

Studies have also demonstrated the feasibility of fabricating micromotors that are fuel-free and powered by external fields. The energy input is then used to create the kinetic driving force for micromotors. Externally powered, which is different to chemical propulsion, can effectively eliminate the disadvantages associated with chemical fuels [4]. The application of a magnetic field has been widely used and examined due to its superior controllability and non-invasiveness, and during the fabrication process, some form of a magnetic component must be included to attain magnetic responsiveness [13, 14]. The application of light has also been investigated as an external power source, and in their study, $Fe_3O_4@Cu_9S_8$ nanoparticles were incorporated into the nanomotor which functioned as the photoactive component and near-infrared-II light was applied to initiate a photothermal effect [15].

The integration of stimulus-responsive hydrogels into micromotors has also been proposed and the notion is that such a combination could extend the applications of micromotors significantly through the addition of new properties such as stimuli-responsive shape morphing [11, 16]. According to the review by Zhou et al., micromotors based on stimuli-responsive hydrogels are typically constructed using actuation-responsive materials, functional components or cargos such as drugs, and stimuli-responsive hydrogels [16]. Actuation-responsive materials can be thought of as the engine that propels the micromotors by responding to chemicals within the physiological environment such as blood or through an externally applied field. Stimuli-responsive hydrogels do not provide any displacement or locomotion action upon exposure to a stimulus environment but can conduct specific tasks once it is exposed to an external stimulus. For instance, stimulus-responsive hydrogels integrated with micromotors would be able to autonomously self-propel or actuated by an external field with precision and perform mechanized minimally invasive operations such as self-folding or targeted delivery of encapsulated cargo upon exposure to an applied stimulus [16].

4.1.2 Soft Robots and Microrobots

Presently, traditional rigid-bodied robots are extensively applied in the manufacturing sector and they can be programmed specifically to carry out a single task efficiently such as spot welding in the automotive industry. However, their adaptability is at times limited [17]. On the other hand, soft robots offer the possibility to connect humans with machines. Soft robots, in contrast to hard-bodied robots, are made from intrinsically stretchable and/or soft materials such as silicone rubber or hydrogels that are inherently compliant and can tolerate deformation and large strains during normal operation [18]. Furthermore, these robots possess a continuously deformable structure with movements similar to human muscle resulting in a relatively high number of degrees of freedom compared to hard-bodied robots [17].

The introduction of hydrogels into the field of soft robotics could widen its utilization in the biomedical arena such as drug delivery [19]. As discussed earlier, stimulus-responsive hydrogels can respond to various external stimuli through the interactions between the external environment and the polymer network. For that reason, their utilization in soft robotics is attractive as they can provide additional capacities through changes in their chemical and physical characteristics when stimulated by external factors such as light, pH, or temperature. Hydrogel-based soft actuators are an example where mechanical motion is manipulated by various applied stimuli [20]. The concept of stimuli-responsiveness on hydrogel was further explored in a recent study where a transformable soft robot was developed by combining cardiac tissue engineering and 3D printing with photosensitive hydrogels and, according to the authors, could be remotely controlled for different motion functions through transformable mechanical structure in response to near-infrared light stimuli [21]. The propulsion system was designed based on the movement of the tail of a swimming whale, and the continuous motion was provided by tissue engineering flexible membrane (cellular engine). For a transition from mobile to stationary status, the floating plane can be optically triggered to retract the wings, which act as an optomechanical "brake" to minimize propulsion output and effectively switch "off" the cellular engine. The authors postulated such an approach could potentially be applied to areas such as targeted and adaptive diagnosis and drug delivery.

It is worth to mention the application of hydrogels in soft robots extends beyond tissue engineering and drug delivery to areas such as actuators, computational circuits, and sensors. A more detailed analysis of hydrogel soft robotics from working mechanisms to future application is provided in the review by Lee et al. [20].

Microrobots, which are small-scale robots ranging from microns to centimeters in size, have attracted significant attention in the biomedical field [19, 22, 23]. It has been hypothesized that the ultimate intention for the development of intelligent microrobots (or even nanorobots) using smart or stimulus-responsive hydrogels is to imitate microcreatures found in nature in an effort for them to adapt to a complex environment [24]. Like micromotors, microrobots can also be controlled or driven by strategies such as chemical reactions, magnetic field, temperature, and light [24]. Presently, due to their small size and motility, microrobots are gaining interest as they could potentially be used in minimally invasive therapy. Moreover, many hydrogel-based microrobots that are biocompatible and biodegradable are also being created. Above all, significant attention has been generated by microrobots that can swell and shrink as a response to changes in temperature created by an alternating magnetic field or external near-infrared stimuli simply because they could be applied as carriers or delivery vehicles for drugs and/or cells [25].

The ability to self-propel is the biggest difference between microrobots and conventional micro- and/or nanodelivery systems. Several challenges are currently being posed by the fabrication of these microrobots such as discovering an efficient way to utilize energies from the surrounding environment as "fuel" for propulsion [4]. More importantly, directional locomotion is critical for hydrogel-based microrobots so that it can travel to specific locations within the human body and perform its intended tasks such as drug release [20].

In addition to the propulsion issue, several problems are also faced by microrobots, which are different to devices and machines constructed at the macroscopic level [4, 20, 26, 27]. Locomotion at small scales is regulated by low Reynolds number and

Brownian motion, and for that reason, the main concern during the design of micro- and nanorobots is the development of engines that are "switched-on" all the time and create sufficient force to overcome the drag forces from the environment such as in a liquid media. Consequently, the design and production of micro- and nanorobots are strongly reliant on active materials that can convert various energy sources into loco- motion continuously [1]. Furthermore, controlling microrobots in a precise manner is also an issue due to Brownian diffusion being greatly increased with a reduc- tion in particle size, which is created from constant collisions between surrounding solvent molecules and particles. Tissues within the human body normally consisted of high-viscosity and flowing fluids, and it is therefore imperative to produce sufficient asymmetric driving forces that can offset the shear stress and high viscous resistance present in the body for the successful application of directional locomotion [20]. Subsequently, asymmetric geometry is typically used to reduce resistance in certain directions; nonetheless, creating enough force to begin the actuation process remains the primary objective in achieving directional locomotion [4, 20].

In general, the motion mechanisms of microrobots are categorized according to the energy source, and they are based either on chemical propulsion or through the application of external physical fields such as light and magnetic fields. Chemically powered microrobots can create bubbles or chemical gradients to propel themselves by triggering chemical reactions in situ in a fluid as a catalyst and/or a reactant [28]. This concept was explored in a study where a hemocompatible magnesium/platinum Janus micromotor could be propelled by hydrogel bubbles generated by the in situ chemical reaction between magnesium and water [29]. In a later study, the group further postulates the magnesium-based micromotors could be utilized as effective vehicles for cells and drugs with proper surface modifications driven by blood plasma or SBF solution (Fig. 4.2) [30]. This was achieved by taking the advantage and capability of poly(N-isopropylacrylamide) hydrogels to uptake and the temperature- controlled release of drugs. In addition, it has been postulated that an enzymeless motor based on phenylboronic acid-modified poly(N-isopropylacrylamide) hydrogel and a surfactant will be able to autonomously travel once it is exposed to glucose molecules [31]. According to the authors, the concept was based on the complexation between phenylboronic acid and glucose causing the swelling of the hydrogel. This swelling then led to the release of the surfactant, and the entire structure was propelled based on the Marangoni effect. Furthermore, they claimed the moving velocity of the motor was directly proportional to the concentration of glucose.

Microrobots propelled by light are often fabricated using photocatalytic or photothermal materials and are thrusted under UV or near-infrared light irradia- tion. Light-driven microrobots possess a significant advantage that the autonomous motion can be controlled remotely via an "on/off" switch or the intensity of the light irradiation [28]. Near-infrared light-driven microrobots are of particular interest for biomedical applications due to the highest transmissivity of near-infrared light by body tissue. A study theorized that near-infrared could be used to control the release of encapsulated cargo such as drugs from the microrobot through the actions of swelling and deswelling [25]. Taking into consideration the rapid response as well as the simple control method of light-propelled hydrogel microrobot, a study attempted

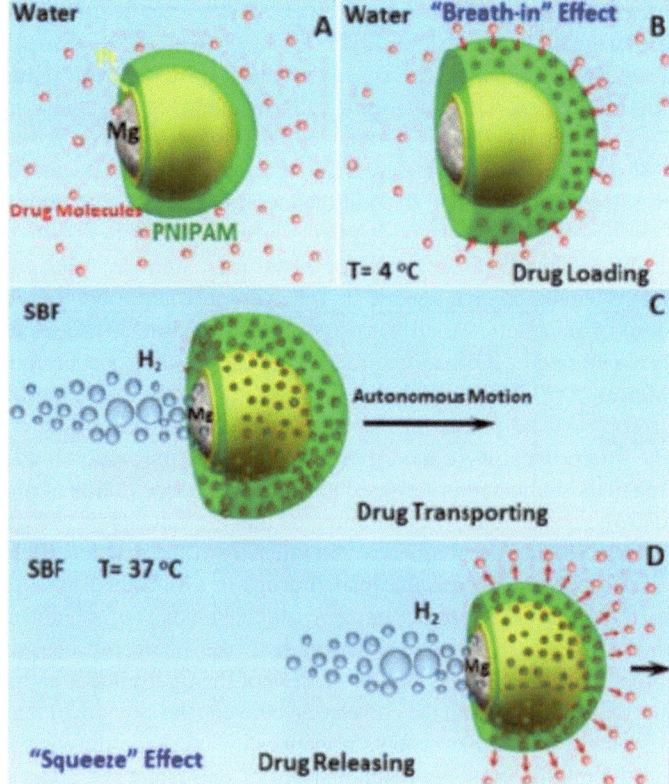

Fig. 4.2 Demonstration of the **a, b** drug loading, **c** transporting, and **d** releasing behaviors of the Mg/Pt PNIPAM Janus micromotors. Reprinted with permission from [30]. Copyright 2014 American Chemical Society

to fabricate light-responsive hydrogel by projection micro-stereolithography-based 3D printing. The hydrogel became responsive to near-infrared light after being decorated with carbon nanoparticles, and hence, the temperature of the hydrogel could be regulated by adjusting the light intensity [22].

Magnetic-propelled microrobots are gaining popularity as they can be actively controlled by changing the direction and intensity of the magnetic field. Subsequently, the microrobots typically contain or carry permanent magnet or magnetic nanoparticles composed of materials such as iron oxide and could be utilized to encapsulate and delivery drugs to targeted sites [32–37]. However, one should keep in mind the potential negative effects provoked by the magnetic nanoparticles that are left behind after completion of drug delivery and the biodegradation of the hydrogel structure [37]. Kim et al. investigated the possibility of fabricating a bilayer hydrogel microrobot consisted of a therapeutic layer and a magnetic nanoparticle layer that could retrieve magnetic nanoparticles after drug delivery. According to their study, the therapeutic layer upon the application of an alternating magnetic field will dissolve to release the encapsulated drugs. After delivery, a magnetic field can also be used to recover the magnetic nanoparticles [37].

Synthetic lethality drugs, in theory, inhibit the growth of typical tumor cells without causing damage to adjacent normal cells, but unfortunately, this is not always the case. Despite normal cells being somewhat affected, it has been postulated that the amount of antitumor drugs delivered to the tumor site reduces, and this gentle inhibition of tumor cells might fail. In addition, increasing the dose of antitumor drugs might also increase the severity of side effects faced by the patients. In an attempt to provide an alternative to resolve some of the issues, a non-target but controllable drug delivery system based on magnetic-driven hydrogel microrobots was utilized to selectively improve synthetic lethality in osteosarcoma in vitro [33]. Gelatin hydrogel microrobots containing Fe_3O_4 nanoparticles were used to transport protein arginine methyltransferase 5 (PRMT5) inhibitors in an in vitro simulated experiment. The loaded microrobots were placed inside a cell culture dish containing a complete medium. The microrobots were then given a directional magnetic field for 5 min, and according to the authors, this was used to simulate the locomotion of microrobots in the blood [33]. Furthermore, they commented that the velocity of 100 μm/s could be achieved by increasing the frequency and intensity of the magnetic field, and the microrobots were able to release the loaded drugs in a sustained manner into the medium for 15–20 min completely. Similarly, another study was attempted to verify the possibility of targeted drug delivery using hydrogel-based capsule microrobots within the blood vessel [35]. The motion process of the microrobots in the vascular microchannel was simulated and the correlation between the strength of the magnetic field and the velocity of the microrobots was investigated. According to the authors, they claimed the microrobots can reach a speed of 800 μm/s but it can go as fast as 3077 μm/s. In addition, the microrobots can continuously climb over a 1000 μm high obstacle under a rotating magnetic field.

As discussed above, controllability and biocompatibility are the main elements that need to be resolved throughout the fabrication of micro- and nanorobots. However, there is one other feature that should also be addressed during the design phase, and it is centered on how we can locate and visualize these synthetic robots in the body. It is crucial to acquire real-time tracking and high-resolution imaging of these micro- and nanorobots if they were to be used in the biomedical arena, especially for in vivo applications [38]. Imaging techniques such as magnetic resonance imaging (MRI) [32, 39] and single-photon emission computed tomography [40] have been explored as possible means to track magnetic microrobots in real-time as well as provide post-operative imaging. It has been suggested that one of the obstacles that is restricting the use of micro- and nanorobots in clinical applications is the integration of imaging techniques with robotics. Addressing issues such as improving imaging resolutions down to the nanoscale, fabricating robots that can be imaged using different imaging techniques, and improvements in contrast to differentiate between surrounding tissues and micro- and nanorobots will expand the applications of micro- and nanorobots in the clinical arena [38, 39].

4.2 Hydrogel Triboelectric Nanogenerators

Triboelectric nanogenerators (TENGs) can convert human biomechanical activities into electricity through the combination of electrostatic induction and triboelectrification [41, 42]. They have been utilized in many healthcare and biomedical scenarios and one such application is their incorporation with drug delivery systems as an innovative approach to attain controllable and on-demand release of drugs [41]. Recently, electrical stimulation therapy has been hypothesized as a therapeutic modality to accelerate wound healing. Based on this notion, studies were carried out to combine TENGs with conductive hydrogels and examine their effect on the rate of wound healing especially for wounds with healing deficiency [43, 44].

It is gradually accepted that the transfer of electrons played a dominant role in contact electrification regardless of the type and state of the triboelectric materials [45, 46]. Four different operation modes of the conventional TENGs have been proposed depending on the direction of the polarization change and electrode configuration: lateral-sliding mode, vertical contact-separation mode, freestanding triboelectric-layer mode, and single-electrode mode (Fig. 4.3) [47].

Rigid materials have always been utilized as electrode materials during the fabrication of conventional TENGs, which limits their ability to withstand large strains as a result of this high rigidity. On the other hand, owing to its biocompatibility, excellent tensile properties, and tunable ionic conductivity, hydrogels have been receiving increasing interest as current collectors in triboelectric nanogenerators [45–47]. Furthermore, by adjusting the microstructure, mechanical property, and conductivity of the hydrogels, many hydrogel-based TENGs have emerged with excellent performance such as achieving a high open circuit voltage of 992 V [48].

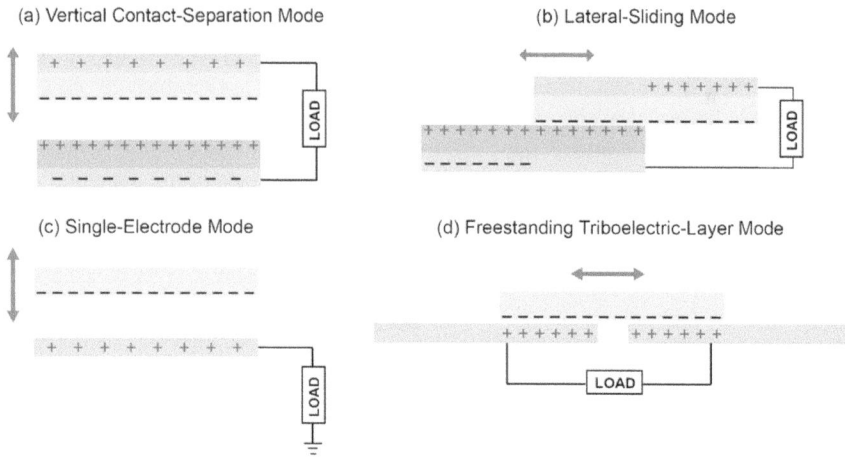

Fig. 4.3 Four working modes of triboelectric nanogenerators. Reprint with permission from [47]

Self-healing can also be achieved which can spontaneously heal within 60 s without any external stimuli [45].

Electrodes are also vital to the integration of TENGs in addition to triboelectric materials. Hydrogels, in comparison to the electronic conductivity displayed by other materials commonly used as electrodes in flexible TENGs such as carbon and sheets and conductive polymer films, possess ionic conductivity that enables optimization and fine-tuning of the charge-carrier density and resistance, as well as the choice of chemical ionic species used in the material [46]. On the other hand, hydrogel electrodes typically have much lower conductivity compared to metal electrodes. The most widely employed configuration when using hydrogels as electrodes in hydrogel-based TENGs is the single-electrode mode based on its simplicity during fabrication. However, dual-electrode TENGs can generate twice the output in comparison to single-electrode TENGs manufactured using the same material and under the same measurement and fabrication conditions [46, 49]. Moreover, the applicability of single-electrode TENGs is restricted since their output performance is dependent on the triboelectric properties of the materials with which they are in contact [49]. Nonetheless, the introduction of conductive additives to the hydrogel can also be used to improve the conductivity when using single-electrode mode such as the addition of MXene nanosheets to hydrogel to create microchannels on the surface to enhance ion transport [50].

Various types of conductivity are demonstrated by hydrogels based on the nature of additives and polymer chains used during the synthesis process, and they can be classified as electron-conductive hydrogels or ion-conductive hydrogels [51]. According to Torres et al., the intention of electron-conductive hydrogels in TENGs is to conduct electrons caused by the differentiated electrostatic charges under the motion of continuous separation and contact through the application of an external load. This is achieved by sandwiching conductive electrodes between two tribo-layers (Fig. 4.4a) [51]. On the other hand, the use of ionic hydrogels in TENG and the generation of electricity is more complex, and the hydrogel is normally covered in an elastomer film and a metal wire is connected [51, 52]. The study by Pu et al. utilized ionic hydrogel as the electrode, and according to their study, electrification occurs once an object contacts the elastomer film at the interface and generates the same quantity of charges with opposite polarities at the surface of the dielectric film and the elastomer, respectively (Fig. 4.4b, <i>) [52]. The static charges on the surface of the elastomer will induce the movement of the ions in the hydrogel to balance the static charges once the two surfaces are separated, forming a layer of excessive ions at the interface (as shown as positive in Fig. 4.4b, <ii>). This will also cause the electrical double layer generated at the metal/electrolyte interface to polarize, forming the same number of negative ions at the interface and positive charges in the metal wire (Fig. 4.4b, <v>). This double layer is achieved when electrons flow from the metal wires to the ground through the external circuits until all the static charges in the elastomer film are screened (Fig. 4.4b, <iii>). The entire process will be reversed if the moving dielectric film travels back to the elastomer film, and an electron flux with the opposite direction will transfer from the ground to the metal/hydrogel interface through the external load (Fig. 4.4b, <iv>). An alternative current will be generated by repeating the separation-contact movement between the dielectric object and the TENG [52]. In another study, an ionic conductive hydrogel was also attempted utilizing graphene oxide and Laponite as the physical crosslinking points, VHB elastomer as the electrification layer, and ionic hydrogel as electrode. According to the authors, the hydrogel displayed high healable capacity and the electrical performance of the TENG could be fully recovered after damage [53].

The application of on-demand drug delivery system enables the control of duration, dosage, and the rate of drug release in addition to enhancing the delivery efficacy

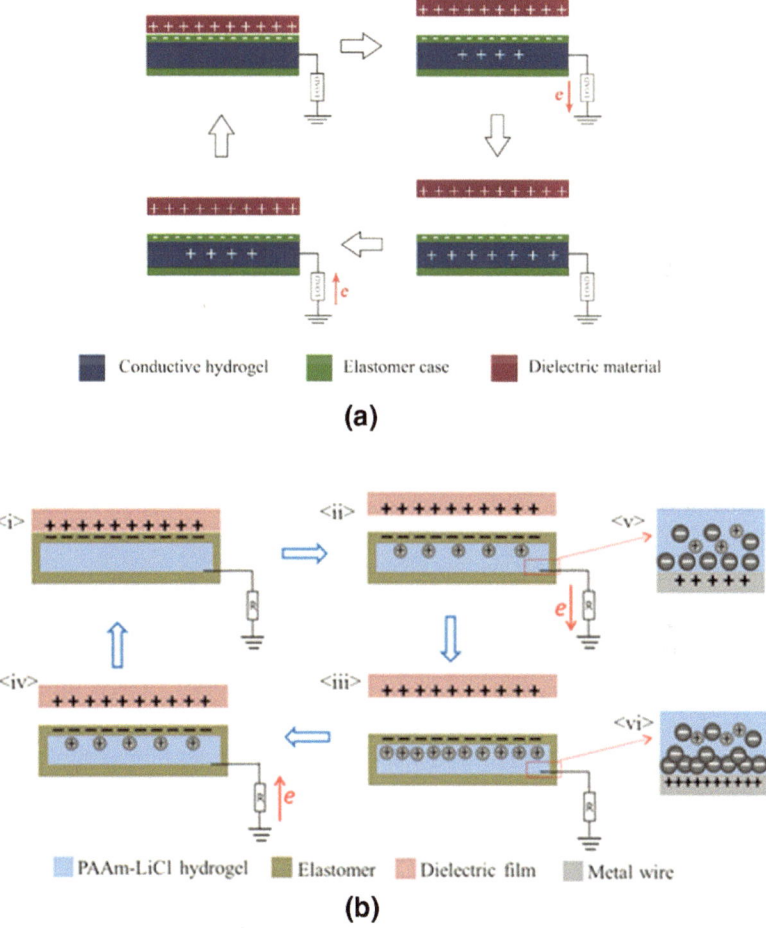

Fig. 4.4 Schematic representation of single-electrode TENGs based on: **a** electron-conductive hydrogels. Reprint with permission from [51]. **b** Ion-conductive hydrogels. In their study, polyacrylamide hydrogel containing lithium chloride (PAAm-LiCl) was used as the ionic hydrogel. Reprint with permission from [52]. http://creativecommons.org/licenses/by/4.0/

during disease development [54]. For the local delivery of drugs, the use of hydrogels as a drug depot especially for situations in which the pathology is localized and contained is an approach worth investigating. As discussed earlier, considerable amount of research has been carried out to develop stimulus-responsive hydrogels such as those responsive to the application of an electric field [55]. In comparison to traditional drug administration such as oral or intravenous administration, a number of advantages are demonstrated by the controlled and local release of drugs such as a finer control over concentration at disease sites and the avoidance of potential off-target side effects since drugs are presented locally [56].

Owing to their unique benefits such as self-administration and convenience, transdermal drug delivery systems have been the focus of clinical research and development [42, 57, 58]. Furthermore, macroscopic drug delivery systems that can self-power using TENGs can address the drawbacks of traditional electrical-regulated systems such as frequent charging or battery replacement [54]. TENG-based self-powered on-demand drug release system converts energy from human biomechanical motions into AC power via triboelectrification and then converts to DC voltage that is suited for use by transdermal patches via a power management circuit [42].

Several theories are utilized by TENG-based system for the delivery of drugs with optimum control and accuracy. As a vital approach of transdermal drug delivery, a high-voltage electrical pulse is needed by electroporation to the skin to create transient pores in plasma membranes which are necessary for the transportation of biomolecules to reach into cells [54, 58]. This notion is often examined in conjunction with sonophoresis for its collaborative outcomes [58].

Iontophoretic drug delivery is the delivery of ionic (charged) drugs into the body using electric current, and this approach has been demonstrated to increase the transdermal permeation of ionic drugs by several magnitudes [59]. Electromigration and electroosmosis are the two key mechanisms of transdermal iontophoretic transport [60, 61]. A continuous low electrical current (< 0.5 m/cm^2) is applied to the skin to initiate the flow of ionic drugs via electromigration, corresponding to the flow of body fluid through electroosmosis to permit the transportation of weakly or neutrally charged molecules [42]. The primary advantage of iontophoresis is its ability to control effectively the rate of drug release by adjusting the electrified time, contact surface area, and the current intensity [42, 58]. This principle was utilized in a study in which a self-powered wearable iontophoretic transdermal drug delivery system was attempted that could be driven and regulated by the energy harvested from biomechanical motions via TENG. A hydrogel-based soft patch with side-by-side electrodes was designed to allow non-invasive iontophoretic transdermal drug delivery [57].

4.3   Concluding Remarks

For any medical micro- and nanorobots in addition to micro- and nanomotors that are intended for in vivo applications, biocompatibility and safety are the main requirements that need to be addressed before their widespread implementation in the clinic. Under ideal conditions, the micro- and nanorobot will conduct their intended tasks such as delivering drugs at a specific location, and once the task has been performed, the robot would either be retrieved through a medical device such as a magnetic field, excreted from the body, or harmlessly reabsorbed in the body through biodegradation. Biodegradation is of vital significance, especially for applications such as cell or drug delivery where micro- and nanorobots would need to perform tasks in hard-to-reach areas of the body such as the brain; hence, any products created during degradation must not pose any risks to the health of the patient. Furthermore, in vivo applications of micro- and nanorobots would require constant and real-time imaging and tracking. To date, imaging techniques such as MRI and single-photon emission computed tomography have been explored for localization of hydrogel-based micro- and nanorobots. Main challenges at present are the insufficient resolution to visualize the micro- and nanorobots and the ability of each medical imaging technique to distinguish individual and large groups of micro- and nanorobots forming a background tissue in real time. Furthermore, the applications of contrast agents are needed for certain imaging techniques to create sufficient contrast to differentiate between various tissues, diseased sites, and the micro- and nanorobots. Subsequently, the retention of these agents could pose long-term cytotoxicity to organs and tissues.

The growing interests into the development and application of hydrogel-based TENGs are due to their ability to harvest biomechanical energy which can then be used to self-power medical devices. Macroscopic drug delivery systems that can self-power using hydrogel-based TENGs can address the drawbacks of traditional electrical-regulated systems such as frequent charging or battery replacement. Furthermore, the use of hydrogels offers the possibility and opportunity for developing wearable devices. Despite the exciting possibilities, more work needs to be done to ascertain the long-term stability of hydrogel-based TENG over longer periods of time. In addition, several challenges also need to be addressed such as stability, output power density, fabrication on a larger scale, and the electronics and electrical components.

References

1. Soto F, Wang J, Ahmed R et al (2020) Medical micro/nanorobots in precision medicine. Adv Sci 7:2002203. https://doi.org/10.1002/advs.202002203
2. Peng F, Tu Y, Wilson DA (2017) Micro/nanomotors towards in vivo application: cell, tissue and biofluid. Chem Soc Rev 46:5289–5310
3. Rastmanesh A, Yaraki MT, Wu J et al (2021) Bioinspired micro/nanomotors towards a self-propelled noninvasive diagnosis and treatment of cancer. Mol Syst Des Eng 6:566–593

4. Li H, Peng F, Yan X et al (2023) Medical micro- and nanomotors in the body. Acta Pharm Sin B 13:517–541
5. Srinivasan A, Roche J, Ravainea V et al (2015) Synthesis of conducting asymmetric hydrogel particles showing autonomous motion. Soft Matter 11:3958–3962
6. Lin XY, Zhu H, Zhao Z et al (2020) Hydrogel-based Janus micromotors capped with functional nanoparticles for environmental applications. Adv Mater Technol 5:2000279. https://doi.org/10.1002/admt.202000279
7. Zhu H, Nawar S, Werner JG et al (2019) Hydrogel micromotors with catalyst-containing liquid core and shell. J Phys Condens Matter 31:214004. https://doi.org/10.1088/1361-648X/ab0822
8. Ariga K, Hill JP, Ji Q (2007) Layer-by-layer assembly as a versatile bottom-up nanofabrication technique for exploratory research and realistic application. Phys Chem Chem Phys 9:2319–2340
9. Lin X, Wu Z, Wu Y et al (2016) Self-propelled micro-/nanomotors based on controlled assembled architectures. Adv Mater 2:1060–1072
10. Yang S, Ren J, Wang H (2022) Injectable micromotor@hydrogel system for antibacterial therapy. Chemistry 28:e202103867. https://doi.org/10.1002/chem.202103867
11. Lin X, Xu B, Zhu H et al (2020) Requirement and development of hydrogel micromotors towards biomedical applications. Research 2020:7659749. https://doi.org/10.34133/2020/7659749
12. Halliwell B, Clement MV, Ramalingam J et al (2000) Hydrogen peroxide. Ubiquitous in cell culture and in vivo? IUBMB Life 50:251–257. https://doi.org/10.1080/713803727
13. Hu N, Wang L, Zhai W et al (2018) Magnetically actuated rolling of star-shaped hydrogel microswimmer. Macromol Chem Phys 219:1700540. https://doi.org/10.1002/macp.201700540
14. Rutkowski S, Si T, Gai M et al (2019) Magnetically-guided hydrogel capsule motors produced via ultrasound assisted hydrodynamic electrospray ionization jetting. J Colloid Interface Sci 541:407–417
15. Wang Y, Chen W, Wang Z et al (2023) NIR-II light powered asymmetric hydrogel nanomotors for enhanced immunochemotherapy. Angew Chem Int Ed Engl 62:e202212866. https://doi.org/10.1002/anie.202212866
16. Zhou H, Dong G, Gao G et al (2022) Hydrogel-based stimuli-responsive micromotors for biomedicine. Cyborg Bionic Syst 2022:9852853. https://doi.org/10.34133/2022/9852853
17. Rus D, Tolley MT (2015) Design, fabrication and control of soft robots. Nature 521:467–475
18. Trivedi D, Rahn CD, Kier WM et al (2008) Soft robotics: biological inspiration, state of the art, and future research. Appl Bionics Biomech 5:99–117
19. Nie M, Zhao Q, Du X (2023) Recent advances in small-scale hydrogel-based robots for adaptive biomedical applications. Nano Res. https://doi.org/10.1007/s12274-023-6184-y
20. Lee Y, Song WJ, Sun JY (2020) Hydrogel soft robotics. Mater Today Phys 15:100258. https://doi.org/10.1016/j.mtphys.2020.100258
21. Xu B, Han X, Hu Y et al (2019) A remotely controlled transformable soft robot based on engineered cardiac tissue construct. Small 15:e1900006. https://doi.org/10.1002/smll.201900006
22. Zhan Z, Chen L, Duan H et al (2021) 3D printed ultra-fast photothermal responsive shape memory hydrogel for microrobots. Int J Extrem Manuf 4:015302. https://doi.org/10.1088/2631-7990/ac376b
23. Kropacek J, Maslen C, van Dijk B et al (2023) Hydrogel microrobots self-assembled into ordered structures with programmable actuation. Adv Intell Syst 5:2300096. https://doi.org/10.1002/aisy.202300096
24. Cao Q, Chen W, Zhong Y et al (2023) Biomedical applications of deformable hydrogel microrobots. Micromachines 14:1824. https://doi.org/10.3390/mi14101824
25. Lee H, Choi H, Lee M et al (2018) Preliminary study on alginate/NIPAM hydrogel-based soft microrobot for controlled drug delivery using electromagnetic actuation and near-infrared stimulus. Biomed Microdevices 20:103. https://doi.org/10.1007/s10544-018-0344-y
26. Sánchez S, Soler L, Katuri J (2015) Chemically powered micro- and nanomotors. Angew Chem Int Ed Engl 54:1414–1444

27. Liu M, Zhao K (2021) Engineering active micro and nanomotors. Micromachines 12:687. https://doi.org/10.3390/mi12060687
28. Xu LL, Mou FZ, Gong HT et al (2017) Light-driven micro/nanomotors: from fundamentals to applications. Chem Soc Rev 46:6905–6926
29. Mou F, Chen C, Ma H et al (2013) Self-propelled micromotors driven by the magnesium-water reaction and their hemolytic properties. Angew Chem Int Ed Engl 52:7208–7212
30. Mou F, Chen C, Zhong Q et al (2014) Autonomous motion and temperature-controlled drug delivery of Mg/Pt-poly(N-isopropylacrylamide) Janus micromotors driven by simulated body fluid and blood plasma. ACS Appl Mater Interfaces 6:9897–9903
31. Li MT, Zhang H, Liu M et al (2017) Motion-based glucose sensing based on a fish-like enzymeless motor. J Mater Chem C 5:4400–4407
32. Lapointe J, Martel S (2009) Thermoresponsive hydrogel with embedded magnetic nanoparticles for the implementation of shrinkable medical microrobots and for targeting and drug delivery applications. Annu Int Conf IEEE Eng Med Biol Soc 2009:4246–4249
33. Mu H, Liu C, Zhang Q et al (2022) Magnetic-driven hydrogel microrobots selectively enhance synthetic lethality in MTAP-deleted osteosarcoma. Front Bioeng Biotechnol 10:911455. https://doi.org/10.3389/fbioe.2022.911455
34. Liu D, Liu X, Chen Z et al (2022) Magnetically driven soft continuum microrobot for intravascular operations in microscale. Cyborg Bionic Syst 2022:9850832. https://doi.org/10.34133/2022/9850832
35. Qiao S, Ouyang H, Zheng X et al (2023) Magnetically actuated hydrogel-based capsule microrobots for intravascular targeted drug delivery. J Mater Chem B 11:6095–6105
36. Li Y, Dong D, Qu Y et al (2023) A multidrug delivery microrobot for the synergistic treatment of cancer. Small 9:e2301889. https://doi.org/10.1002/smll.202301889
37. Kim DI, Lee H, Kwon SH et al (2020) Bilayer hydrogel sheet-type intraocular microrobot for drug delivery and magnetic nanoparticles retrieval. Adv Healthc Mater 9:e2000118. https://doi.org/10.1002/adhm.202000118
38. Wang B, Zhang Y, Zhang L (2018) Recent progress on micro- and nano-robots: towards in vivo tracking and localization. Quant Imaging Med Surg 8:461–479
39. Go G, Yoo A, Nguyen KT et al (2022) Multifunctional microrobot with real-time visualization and magnetic resonance imaging for chemoembolization therapy of liver cancer. Sci Adv 8:eabq8545. https://doi.org/10.1126/sciadv.abq8545
40. Iacovacci V, Blanc A, Huang H et al (2019) High-resolution SPECT imaging of stimuli-responsive soft microrobots. Small 15:e1900709. https://doi.org/10.1002/smll.201900709
41. Li X, Tat T, Chen J (2021) Triboelectric nanogenerators for self-powered drug delivery. Trends Chem 3:765–778
42. Ikram M, Mahmud MAP (2023) Advanced triboelectric nanogenerator-driven drug delivery systems for targeted therapies. Drug Deliv Transl Res 13:54–78
43. Jeong SH, Lee Y, Lee MG et al (2021) Accelerated wound healing with an ionic patch assisted by a triboelectric nanogenerator. Nano Energy 79:105463. https://doi.org/10.1016/j.nanoen.2020.105463
44. Sharma A, Panwar V, Mondal B et al (2022) Electrical stimulation induced by a piezo-driven triboelectric nanogenerator and electroactive hydrogel composite, accelerate wound repair. Nano Energy 99:107419. https://doi.org/10.1016/j.nanoen.2022.107419
45. Wang SJ, Jing X, Mi HY et al (2022) Development and applications of hydrogel-based triboelectric nanogenerators: a mini-review. Polymers 14:1452. https://doi.org/10.3390/polym14071452
46. Wu Y, Luo Y, Cuthbert TJ et al (2022) Hydrogels as soft ionic conductors in flexible and wearable triboelectric nanogenerators. Adv Sci 9:e2106008. https://doi.org/10.1002/advs.202106008
47. Wu C, Wang AC, Ding W et al (2019) Triboelectric nanogenerator: a foundation of the energy for the new era. Adv Energy Mater 9:1802906. https://doi.org/10.1002/aenm.201802906
48. Li G, Zhang J, Huang F et al (2021) Transparent, stretchable and high-performance triboelectric nanogenerator based on dehydration-free ionically conductive solid polymer electrode. Nano Energy 88:106289. https://doi.org/10.1016/j.nanoen.2021.106289

49. Wang Z, Li N, Zhang Z et al (2023) Hydrogel-based energy harvesters and self-powered sensors for wearable applications. Nanoenergy Adv 3:315–342

50. Luo X, Zhu L, Wang YC et al (2021) A flexible multifunctional triboelectric nanogenerator based on MXene/PVA hydrogel. Adv Funct Mater 31:2104928. https://doi.org/10.1002/adfm.202104928

51. Torres FG, Troncoso OP, De-la-Torre GE (2022) Hydrogel-based triboelectric nanogenerators: properties, performance, and applications. Int J Energy Res 46:5603–5624

52. Pu X, Liu M, Chen X et al (2017) Ultrastretchable, transparent triboelectric nanogenerator as electronic skin for biomechanical energy harvesting and tactile sensing. Sci Adv 3:e1700015. https://doi.org/10.1126/sciadv.1700015

53. Li G, Li L, Zhang P et al (2021) Ultra-stretchable and healable hydrogel-based triboelectric nanogenerators for energy harvesting and self-powered sensing. RSC Adv 11:17437–17444

54. Liu Z, Li L (2021) Self-powered drug-delivery systems based on triboelectric nanogenerator. Adv Energy Sustain Res 2:2100013. https://doi.org/10.1002/aesr.202100013

55. Oliva N, Conde J, Wang K et al (2017) Designing hydrogels for on-demand therapy. Acc Chem Res 50:669–679

56. Brudno Y, Mooney DJ (2015) On-demand drug delivery from local depots. J Control Release 219:8–17

57. Wu C, Jiang P, Li W et al (2020) Self-powered iontophoretic transdermal drug delivery system driven and regulated by biomechanical motions. Adv Funct Mater 30:1907378. https://doi.org/10.1002/adfm.201907378

58. Adhikary P, Mahmud MAP, Solaiman T et al (2022) Recent advances on biomechanical motion-driven triboelectric nanogenerators for drug delivery. Nano Today 45:101513. https://doi.org/10.1016/j.nantod.2022.101513

59. Banga AK, Chien YW (1988) Iontophoretic delivery of drugs: fundamentals, developments and biomedical applications. J Control Release 7:1–24

60. Kalia YN, Naik A, Garrison J et al (2004) Iontophoretic drug delivery. Adv Drug Deliv Rev 56:619–658

61. Ita K (2016) Transdermal iontophoretic drug delivery: advances and challenges. J Drug Target 24:386–391